Regulating bodies

Professor Turner's book provides a framework for the development of a new sub-field, namely the sociology of the body. Through an examination of various philosophical traditions (phenomenology, philosophical anthropology, structuralism and postmodernism) the book shows how the human body has been ignored or neglected by mainstream social theory. In attempting to integrate these different traditions, Professor Turner demonstrates how the absent body has impoverished not only the sociology of health and illness but the very foundations of sociology itself. There are three major aspects to this argument. First, it is impossible to develop an adequate theory of social action without a conception of the embodied social agent. Second, the idea of embodiment offers a fundamental critique of the positivistic side of the medical model of illness, and thus offers a new theoretical basis for medical sociology. Third, following the work of Michel Foucault, Turner demonstrates that medical practice functions as a moral discourse which produces a regulation of the body. By providing a general account of the problem of the body in modern society, this study, building on Professor Turner's previous studies of *The Body and Society* (1984) and *Medical Power and Social Knowledge* (1987), attempts to solve many of the existing epistemological and theoretical difficulties in social theories of the body.

Professor Turner has provided a major synthesis of his earlier work on the sociology of the body, established the idea of embodiment as fundamental to the sociology of health and illness, and pointed the way forward to new areas of cultural analysis. This volume is a major university text for sociology, philosophy and feminist theory.

Bryan Turner is Professor of Sociology at Essex University.

Regulating bodies

Essays in medical sociology

Bryan S. Turner

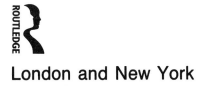

London and New York

First published in 1992
by Routledge
11 New Fetter Lane, London EC4P 4EE

Simultaneously published in the USA and Canada
by Routledge
a division of Routledge, Chapman and Hall Inc.
29 West 35th Street, New York, NY 10001

© 1992 Bryan S. Turner

Phototypeset in 10pt Baskerville by
Mews Photosetting, Beckenham, Kent
Printed and bound in Great Britain by
Biddles Ltd, Guildford and King's Lynn

British Library Cataloguing in Publication Data
A catalogue record for this book is available from the British Library

Library of Congress Cataloging in Publication Data
A catalog record for this book is available from the Library of Congress

ISBN 0-415-06963-7 (hbk)
ISBN 0-415-08264-1 (pbk)

Contents

To Adelaide, Joan, Karen and Mathilde

Acknowledgements

Chris Rojek has been particularly important and supportive in helping me to bring this collection of lectures and essays together as a coherent discussion of the problems of bodily regulation in relation to medical control and surveillance. I would also like to thank Dr Richard Fardon of the London School of Oriental and African Studies for taking the care to read my work on the body with such critical scrutiny. Dr Fardon is completing an intellectual biography of Mary Douglas and he is thus ideally suited to interrogate my own contribution to the study of the body. His searching interview, which took place at Routledge's offices in London on 11 July 1991, has forced me to think deeply and critically about the project of a sociology of the body. A similar debt should also be acknowledged with respect to Arthur Frank, University of Calgary, who has on several occasions forced me to reanalyse my approach to the questions of the body and embodiment.

Over a longer period Mike Featherstone has drawn me towards and carefully guided me through a wide range of modern thinkers whose work, in one way or another, has addressed questions about the body, embodiment, biology and civilization: Michel Foucault, Jean Baudrillard, Pierre Bourdieu and Norbert Elias. The breadth of his interests, which are represented in the editorial work of *Theory, Culture & Society*, has been a source of inspiration.

Georg Stauth, currently at the University of Singapore, forced me in ways which as yet are obscure to work my way through philosophical anthropology, Nietzsche, Ludwig Klages, Kierkegaard and Scheler towards a broader understanding of social theory in Germany.

It has become obvious in putting these lectures and essays together that my work reflects an as yet incomplete synthesis of medical sociology and the sociology of religion. My original interest in this area of sociology and especially with reference to the concept of theodicy was formed by Roland Robertson and Trevor Ling at the University of Leeds in the early 1960s.

Finally, Ken Plummer of the University of Essex provided me with a number of very handy references to the work of G.H. Mead and symbolic interaction, which helped me to think my way through a number of problems in the final stages of this book.

Chapter 2 was originally a public lecture given at the American–German theory group conference at the University of Maryland in 1990 on structuration theory. I am grateful to the organizer George Ritzer and participants for their critical comments. Chapter 4 was originally published in *Sociology of Health & Illness* 1990, vol. 12(1):1–23. Chapter 6 was published in the *British Journal of Sociology* 1982, vol. 33(22):254–69. I am grateful for permission to reprint these articles. Chapter 7 was first published in *The Sociological Review* 1990, vol. 38(1):1–18. Chapter 8 was published in *The Australian and New Zealand Journal of Sociology* 1990, vol. 26(2):157–69. I am grateful to Basil Blackwells for permission to reprint Chapters 4 and 7, and to *The Australian and New Zealand Journal of Sociology* for permission to reprint Chapter 8. Chapter 5 was the keynote address to the British Sociological Association medical sociology annual conference in 1989. Chapter 3 was inspired by the work of Robert Hertz and by the belief that the hand is the most beautiful part of the body. I would like to thank Chris Rojek for organizing the discussion which forms the basis of the Conclusion.

Author's preface
Towards the somatic society

Autobiographical details about the development of ideas and their publication may no longer be regarded in some quarters as authoritative according to the norms of contemporary textual analysis, but in my view they may at least count as partial evidence. How can one know where, when and why a particular train of thought eventually appears to result in 'a project'? My own development over the last ten years seems in any event deeply bound up with the journal *Theory, Culture & Society*, and so 'my' project is inevitably collective. The diversity of interests which is represented by that journal may go some way to explaining my own somewhat eclectic approach to sociological theory. Much of the work of the journal over the last decade has been associated with the study of consumerism (Featherstone 1991), leisure, sport and the body (Featherstone *et al.* 1991). Partly as a consequence of this array of topics, my attempt to develop a sociology of the body has been shaped by a plethora of writers: Berger, Bourdieu, Deleuze, Douglas, Elias, Foucault and O'Neill. If there is any epistemological theme in my sociological work, it is based on a hostility to intellectual specialization which is the mark of the professional academic. My heroes have always worked on a very broad canvas.

To be more specific and to start somewhere near the beginning, my interest in the sociology of the body was the consequence of a number of diverse intellectual issues and concerns. It grew partly out of studying Peter Berger's interpretation of the legacy of Marx, Weber and Durkheim for the sociological understanding of how 'the world' is socially constructed (Berger 1969). It was by approaching the sociology of religion through the perspective of Berger's sociology of knowledge that I came eventually to see the body as the key to debates about theodicy, soteriology and meaning in the Abrahamic religions. Because my own academic research had been originally in the sociology of religion, I was drawn especially to the concept of theodicy as a way of understanding the tensions between our embodiment and the requirements of sacred cultures (Turner 1981). In particular it brought me to conceptualize the structure of 'the world' in terms of Weber's analysis of religious orientations by reference to three elements: politics as violence versus the ethic of (in Weber's terms) 'brotherly love'; sexuality as destructive eros versus self-giving agape; and economics as rational self-interest versus the sacred calling of charity. It was in Berger's social constructionist approach to the sacred canopy that these themes had converged around the problem of meaning.

However, it was not until much later that I realized that Berger's own account of the nature of social institutions was a contribution to a tradition of philosophical anthropology, which was closely associated with the work of Helmuth Plessner and Arnold Gehlen. Since Gehlen was fundamentally influenced by Nietzsche's idea that 'Man' is an unfinished animal, it was almost inevitable that these interests in Weber, Berger and Gehlen should force me into a detailed study of Nietzsche. In *Nietzsche's Dance* (Stauth and Turner 1988), we attempted to uncover the body behind Nietzsche's celebration of everyday life. The title of course refers to Nietzsche's humorous belief that, while previous German philosophers had slowly stumbled towards the truth after enormous intellectual labour, he danced his way joyously towards life. Reading Nietzsche intensively in the early 1980s convinced me that ideas about the body and health were the key to his general philosophy.

As a schoolboy in Birmingham in the 1950s, I had been taught by the great Marxist philosopher Alfred Sohn-Rethel, but it was not until late in my academic career that I became familiar with his *Geistige und Körperliche Arbeit* (1970). Interestingly, the English translation of this contrast ('intellectual and manual labour') fails to capture the literal German contrast beween 'spirit' and 'body'. If *körperliche Arbeit* might be literally translated as 'bodily labour', then my own sociological interests can be seen as a contribution to the understanding of our labour on the body as well as into bodily labours.

In retrospect, it was not until I spent a year in Germany (at Freiburg and Bielefeld) as an Alexander von Humboldt Fellow in 1987–88 that the impact of German social philosophy (Marx, Nietzsche, Weber, Troeltsch, Gehlen and Sohn-Rethel) began to condense into a more coherent perspective. This growing interest in German social theory continued while I was professor of general social sciences at the university of Utrecht (1988–90) where there was a keen interest in the works of Mannheim, Scheler, Habermas, Apel and others. I began to appreciate more fully the long-established contrast in German thought between the immediacy, practicality and sensuality of the life-world and the regimentation, externality and constraints of the institutional structure of the social system. This contrast was not only fundamental to Nietzsche but it also runs through Weber, Heidegger, Gehlen, Heller and Habermas. Of course, the ways in which these various writers have managed and developed that contrast have gone in very different directions. However, these writers are closely engaged in the development of a philosophical anthropology of everyday life (Heller 1984). It is not surprising, given the legacy of the Bismarckian civil service and the culture of the *Obrigkeitsstaat*, that German social theorists should be concerned to understand the impact of state structures on everyday life. We find in Weber in particular this fear of the inescapable impact of bureaucracy on the individual, who is merely a cog in the machine.

In order to comprehend this everyday world, or life-world, it appears to me that a sociology of the body is a necessary condition for understanding everyday routines, conditions and requirements. Everyday life is about the production and reproduction of bodies; we have to grasp this elementary fact before we can go on to talk about the production of 'the person' (Heller 1984:51). This approach is one way of understanding the idea of the 'dull compulsion of economic relations' (Marx 1970, vol. 1:737), which preoccupied much of my work with Nicholas Abercrombie and Stephen Hill in writing *The Dominant Ideology Thesis* (Abercrombie *et al.* 1980). From the perspective of a Marxist philosophical anthropology, I would prefer now to talk about the 'dull compulsion of everyday life and bodily practices', because I see the routine social maintenance of the body as the foundation of this dull compulsion.

By becoming immersed in this literature on philosophical anthropology via the work of the young Marx, Nietzsche and Gehlen, I began to see that Weber, far from being engaged in a narrow debate with Marx over capitalism, was in fact concerned to understand how civilizational institutions shaped human personality in such a way that individual autonomy was challenged by the rationalization process. I tried to explore some of these themes (with Georg Stauth) in *Nietzsche's Dance* (Stauth and Turner 1988) and in *Max Weber, from History to Modernity* (Turner 1992). These issues have been treated with enormous success by Wilhelm Hennis in his *Max Weber, Essays in Reconstruction* (Hennis 1988). It is therefore not surprising that, when I began seriously to confront the work of Michel Foucault, which in turn was a consequence of following Nietzsche's impact on French social theory, that I should see Foucault through the eyes of a Nietzschean Weber. For me, Foucault's account of panopticism, the carceral society, disciplines and regulative practices read like a version of the processes of rationalization. In the late 1970s, I also came to the conclusion that Norbert Elias's concept of the civilizing process (Elias 1978) was also a version of the rationalization of the body through table manners, courtly norms of good conduct, etiquette, decorum and so forth. While the comparison of rationalization and civilization processes is often made in order to note the relationship between Weber and Elias (Bogner 1989), it was Alfred Weber, Max Weber's brother, who in fact had a much more developed view of the relationships between socialization, civilization and rationalization (Weber 1920-1). Alfred Weber's whole concept of 'cultural sociology' was based on an historical understanding of the contrast between culture and civilization, where civilization is about the spread of technology. Thus, while I do not think that Elias's theory of civilizing processes in terms of human cultivation was so original in its context, Elias, in addition to his historical analysis of the civilization of the body, also had an important understanding of the complex relationship between self and body (Elias 1991: 188-9).

If we were to summarize Weber's sociology, we might do so under the headings of power, personality and discipline. What are the social processes that produce the disciplined personality? I have suggested that we cannot understand the disciplined mind independently of the disciplined body. This insight was fundamental to Foucault's work on the training of the self. I had (with Mike Hepworth) attempted to understand some aspects of this relationship through a study of confession (Hepworth and Turner 1982). Thus, it was merely a small step towards seeing the body as the key to these very diverse and different intellectual currents. The history of the West was not so much the transformation of culture under the impact of rationality as the transformation of the human body via a myriad of practices. Medicalization, secularization and rationalization appeared to be the great forces which had operated on the body, or more precisely on the body-in-the-everyday-world.

Thus, the publication of *The Body and Society* in 1984 brought to a conclusion a period in my own intellectual development, which had been heavily influenced by the sociologies of Karl Marx and Max Weber. British social theory in the 1970s had been significantly engaged with debates about base and superstructure, modes of production and dominant ideologies. The book reflected, therefore, a complex combination of 'materialisms' which I had derived from Marxism via a critical reading of Louis Althusser, and a growing appreciation of French social theory, especially the work of Michel Foucault. Against structuralism, however, I had attempted to retain from the early Marx a concern for 'sensualism' which allowed me to understand 'materialism' in terms of human embodiment and embodied practices. In the 1970s, my interest in Ludwig Feuerbach did, of course, look distinctively odd. At the same time, Weber's idea about the processes of rationalization in society provided a perspective on the impact of science and regulation on human bodies in the development of modern societies, and this led to an interest in medicine as the key institution in this secular regulation of bodies.

I have often been criticized as a result for eclecticism, and for a lack of theoretical integration. This criticism arises partly because of the diversity of topics which I have approached in religion, theodicy, materialism and the body in *Religion and Social Theory* (1983), the body, medicine and feminism in *The Body and Society* (1984), and sociological theory, the body and health in *Medical Power and Social Knowledge* (1987). This apparent eclecticism may also reflect the very different conditions and circumstances under which these various books were written. *Religion and Social Theory* was written at the university of Aberdeen, where I had been working systematically on Nietzsche, Weber and Foucault. It was also in Aberdeen university library that I was fortunate enough to study the works of George Cheyne in the original first editions; this study was the background to the work on dietetics. *The Body and Society* was written in Australia at Flinders university, where I had become especially engaged with feminist debates about patriarchal medicine. *Nietzsche's Dance* was written with Georg Stauth in Australia and Germany, when we were both interested in Ludwig Klages and the Stefan George Circle. Similarly *Medical Power and Social*

Knowledge, where the influence of Foucault is the most obvious and prominent, was started in Australia and completed in Germany. In England, I have recently edited (with Featherstone and Hepworth) *The Body* (1991). These geographical and institutional shifts have promoted a diversity of interests, but there are also important themes running throughout these studies. However, in *Regulating Bodies*, there is a more specific empirical focus, namely the body in relation to major issues in medical sociology. I have also attempted to provide a more self-conscious intellectual justification for my diverse analytical and empirical concerns.

Clearly a Marxist problematic was relatively dominant in *The Body and Society*. It attempted to understand the importance of a sociology of the body in terms of changes in the mode of production, namely in the shift from feudalism to capitalism. I took for granted the historical arguments about the transition from feudalism to capitalism. The thesis was that we could understand the disciplinary requirements of a capitalist civilization in terms of the ascetic practices of the body. The necessity to regulate labour provided a materialist and historical perspective on the development of disciplines. The dietary regime appeared to be a perfect way into this analysis, partly because the term 'regime' permits one to connect a medical life-plan with the larger notion of a government.

Almost a decade later, these arguments look somewhat dated. The intellectual edifice of Marxism as a theory of history and social organization has been shaken by the fall of communist regimes in eastern Europe and the Soviet Union. There is some agreement that we are already intellectually in a period of post-Marxism. By comparison with the orthodox rigidities of the late 1960s and early 1970s, contemporary social theory in general and sociology in particular are, possibly as a consequence of poststructuralism and postmodernism, more open, diverse and interdisciplinary. The interpretation of Weberian sociology has gone through a profound revolution in which Weber is seen to be a Nietzschean theorist of the crisis of modernity rather than narrowly interpreted as an analyst of industrial capitalism; the Nietzschean roots of Weber's sociology of modern mentalities are now regarded as decisive in the formation of Weber's sociological perspective. Michel Foucault, who had influenced my view of 'the government of the body' in terms of medical regimes and who as a result shaped the arguments about knowledge/power in *Medical Power and Social Knowledge*, died in

June 1984. Now that the full scope and scale of his work can be more adequately appreciated and understood, my original use of his approach appears in retrospect to be inadequate. I hope that this new study of 'the regulation of bodies' reflects these new developments both in society and social theory.

As my work on the sociology of the body has evolved I have attempted to move away from the original Marxist anchorage to develop a general orientation, which will debate and include recent intellectual movements in poststructuralism, feminism, literary theory and postmodernism. However, the intellectual targets which I have set myself are not the easiest to reach. My starting point is with the basic idea of sociology, namely the characteristics of social action. In this sense, my concern for a sociology of the body follows from a dissatisfaction with the treatment of rationality, agency and agent in Weber's sociology of action. It is well known that Weber was unable to solve the problem of the relationship between rational actions, affective action, the non-rational, and the symbolic in his methodology of the social sciences (Sica 1988). The economic idea of utility could not as an ideal type be easily extended to social action as such. Talcott Parsons wrestled with a similar problem in his attempt to develop a voluntaristic theory of action (Gould 1991). Although the social theory of Anthony Giddens had done much to transform the legacy of agency and structure, I did not feel that he had done enough to theorize the body as a starting point for the theory of action. He has on a number of occasions recognized the importance of the body (Giddens 1984:36) and yet his social actor was still primarily a knowing and choosing actor, engaging in a reflective and rational appraisal of the conditions of action. Giddens had transformed the Weberian actor in terms of a reading of Heidegger, but the problems of embodied agency for a theory of structuration had hardly been raised, let alone resolved. I do not feel that Giddens's recent interest in a sociology of the self has fully addressed the question of the embodied self (Giddens 1991). Giddens wants to treat reflexiveness as the central feature of modernity and hence he is interested in the idea of the body as a topic of reflexivity. While this approach is perfectly legitimate and important, I am still convinced that Giddens's 'social actor' is an implicitly disembodied consciousness. By contrast, I tend to agree with Dennis Wrong that, against the 'oversocialized man' in modern sociology, we have to assert that in the beginning was the body (Wrong 1961). To understand social action, we need some conception of 'the social actor' and in particular we need a conception of the embodied actor which will transcend the all-pervasive Cartesian division of mind–body. To achieve this goal we have to develop a phenomenology of 'the lived body', which in my view has to be derived from a diverse set of traditions: life-philosophy, philosophical anthropology and phenomenology (Honneth and Joas 1988). However, in this study I attempt to combine this tradition with an appreciation of the body as representation. This task is difficult because it appears to fly in the face of recent philosophical trends which separate out the philosophy of experience from the philosophy of knowledge. Indeed, this division between a philosophy of subjective experience and a philosophy of the structures of knowledge was fundamental to Foucault's project (Foucault 1985). *Regulating Bodies* attempts to cut across that division.

The idea of the body as representation, and in particular as a representation of the fundamental features of society, is not a recent development. Anthropologists, especially through the influence of Mary Douglas, have studied the body as a narrative of social processes and social structures. However, in contemporary social theory, as a consequence of the growing interest in 'deconstructionism' in the work of Paul de Man and Jacques Derrida, it is fashionable to regard the body as a text, or as the effect of a discourse. The body is socially constructed through discourses – medical, moral, artistic, commercial. These ideas have a number of consequences. They problematize 'the body' and make any essentialist notion of the body as a living organism, which could be studied through the neutral gaze of science, difficult to sustain. Social constructionism – the view that the body is fabricated by scientific discourses in medicine – calls into question the claims of expert knowledge. These deconstructive, critical readings of the body – for which I have a great sympathy – are normally regarded as 'anti-humanist', because they challenge the modernist understanding of knowledge, subjectivity and the subject–object relationship. The Text replaces the knowing, conscious Author; knowledge is an effect of textuality not subjective comprehension. In specific terms, structuralism and poststructuralism in France were an attack on the existentialism of Sartre and the phenomenology of Merleau-Ponty. The problem which dominates *Regulating Bodies* is how to comprehend the historical evolution of the discourse about the body, to acknowledge how our perspectives on the body are the product of social constructions, and to retain an appreciation for the phenomenological nature of the lived body.

This attempt to find some integration between structuralism, poststructuralism and phenomenology is specifically considered in Chapters 1, 3 and 9. I adopt a number of separate theoretical moves to achieve this analytic integration. One crucial feature of this defence of theoretical diversity is based on a distinction in phenomenology between the difference in the German language between the body as *Körper* and the body as *Leib,* that is between the objective-instrumental body and the subjective-animate body. Although deconstructionist readings of cultural representations of the body can produce significant results, it is difficult wholly to reject the facticity of the instrumental-objective body. To take an almost random example, the uncorrelated asymmetries of the human body (*situs,* handedness, hand-clasping, cerebral dominance and arm-folding) are very general features of the human organism, some of which are shared with other vertebrates (Wolpert 1991). However, it is important to distinguish between the asymmetry of the objective body (*Körper*) and the subjective experiences of the asymmetries of the body as *Leib.* It is equally important to distinguish between organic asymmetry (handedness) and the cultural representations and social meanings of right-handedness. One can, therefore, think of various layers or levels of analysis depending on the issues involved and the problems which scientists set themselves.

In this author's preface, I do not intend to elaborate these arguments. The complexity of the hand as an object of analysis is treated at some length in Chapter 3. The point of this comment is to claim that *Regulating Bodies* offers a more self-conscious elaboration of the epistemological difficulties of talking about the body than I was able to provide in *The Body and Society.* I have simply become more conscious of the underlying problems of 'the body' than was the case in my earlier work. In this volume, I attempt to direct attention to specific issues in medicine, medical knowledge and medical sociology.

There are, therefore, some important changes between *The Body and Society* and *Regulating Bodies.* The Marxist influence of my earlier work has largely disappeared, but the Weberian interest in medicalization (the transformation of general social problems into technical medical concerns, and the elaboration of medicine as the basis of social control) as a specific form of rationalization remains a common theme. The attempt to engage with feminism through the development of an alternative to the theory of patriarchy, namely a concept of patrism, is somewhat replaced in this new study by a more general critique of rationalistic models of human behaviour. The concern to understand the complex relationship between religion and medicine in *The Body and Society* is far less prominent in *Regulating Bodies*. The somewhat rigid discussion of the Hobbesian problem of order as a method of approaching the idea of body regulations (Turner 1984:90) has been replaced in this study by a more general interest in the body in relation to theories of social action.

However, one continuity in these publications is in fact a conviction that the question of the body continues to dominate politics and culture in the late twentieth century. Much of my work has been inspired by Foucault's argument that 'The disciplines of the body and the regulations of the population constitute the two poles around which the organization of power over life was deployed' (Foucault 1981:139). This bio-politics lies at the core of modern systems of power. Where then is resistance? What form does it assume? In *The Body and Society*, and more specifically in *Nietzsche's Dance* (Stauth and Turner 1988), the body, following Nietzsche and Foucault, was conceived as a site of resistance, as a source of playful energies and as the Dionysian principle. The body sets up resistances to the rational processes of standardization, regulation and control. Foucault was especially concerned, therefore, to understand the nature of those practices which, with the emergence of capitalism, produced the docile body, the disciplined body, and the productive body. He conceptualized the emergence of a new type of society, the carceral (Foucault 1979).

These Foucauldian ideas, which to my mind are parallel to Weberian categories of rationalization and disenchantment, have shaped my work since my entry into this field in the early 1980s with 'The government of the body' (Turner 1982). Perhaps the next stage in the development of these arguments should be to detect a change in the status of the body in advanced capitalist societies. Foucault was concerned to understand how the body was harnessed to the needs of an emerging capitalist society and to the bureaucratic state. Because traditional forms of leisure and activity had to be transformed, there was a new emphasis on the productive and disciplined body; innovative schemes of body regulation emerged in factories, schools and prisons. Panopticism released new corporeal powers and energies to drive society to new levels of productivity.

In the advanced industrial societies, we can detect a transform-ation, because the body in the late twentieth century is no longer the productive body. Whereas most social scientists are concerned to understand leisure as a mechanism of body production in consumer society, it seems to me that the body is being increasingly experi-enced, discussed and represented as a limit and as a brake on growth. The most obvious illustration of the body as the limiting horizon of advanced capitalism is the question of ageing popula-tions and the so-called 'burden of dependency'. Fixed patterns of age-related retirement and ageing populations mean that around 14 per cent of national income is absorbed by pensions. Throughout the second half of the twentieth century, the white populations of northern Europe, such as Germany and Finland, have ceased to reproduce themselves. We might regard this new anxiety about dependency, ageing, retirement, the social consequences of Alzheimer's disease, and the failure of population replacement as a form of hyper-Malthusianism. These anxieties are fuelled in particular by the growth of HIV and AIDS in the heterosexual population, especially in the ethnically diverse, underprivileged, intercity areas of North America. These new diseases have major implications for future welfare programmes. In the developing world, AIDS may, in societies like India, preclude any real oppor-tunity for economic growth. However, they may also be used as a basis for an extended medical surveillance of human populations.

While Marx was obviously aware of the complex interaction between men and machines, he could not envisage how in every-day life we interact with answering machines, communicate by fax, tape-record our intimate thoughts, and video our surgical opera-tions. Perhaps in any case the human body can no longer be the carrier of labour-power. The body is no longer sufficiently efficient to achieve the goals of modern production methods. Computers and cyborgs will have to supplement the human body in both economics and warfare. The 'clean war' against Iraq is probably the model of future global struggles, at least between the most advanced social systems. In future wars, one can anticipate a growing dependence on computerized bodies or cyborgs, because the human body is no longer sufficiently reliable, efficient and effective. These changes indicate that regulating bodies will continue to be a fundamental activity of political and social life.

These global developments suggest a new concept for the bio-politics of the twenty-first century, namely the development of the somatic society. Many new terms have been coined recently to express the character of modern societies: postindustrial, postfordist, postmodern, or semiotic society. The earlier ideas of the leisure society, the consumer society, or the postindustrial society expressed a certain optimism or confidence about the future. These concepts have been replaced by a more nervous paradigm of disorganization, especially in the neo-Marxist view of disorganized capitalism or the postmodern vision of the hyper-real society. There is a new awareness of risk in social relations, especially in sexual relationships where the gamble on health is part of a pornographic thrill in the chance encounter.

The body is obviously very important in these secular eschatologies. In postmodern debates, the body acquires the aura of a special nostalgia in a world of risk, uncertainty and disorganization. The body is a significant feature of hyper-real America in Jean Baudrillard's *America* (1988). It is an important dimension to the panic culture of postmodern sensibilities (Kroker and Kroker 1987). Our metaphors of disorder perhaps reflect our consciousness that death visits our bodies, not through acts of overt violence, but secretly through cancerous growths, silent viruses and humiliating strokes. These medical and demographic developments, therefore, lend weight to the need for a new concept of modern societies as somatic. We might define the somatic society as a social system in which the body, as simultaneously constraint and resistance, is the principal field of political and cultural activity. The body is the dominant means by which the tensions and crises of society are thematized; the body provides the stuff of our ideological reflections on the nature of our unpredictable time. We live in a world which is out of joint. The feminist movement, pensioners' lobbies, AIDS campaigns, pro- and anti-abortion cases, fertility and infertility programmes, institutions to store human organs, safe-sex campaigns, global sporting spectaculars, movements for preventive medicine, campaigns to control global tourist pornography, and various aspects of the Green Movement are all major aspects of the bio-politics of the somatic society. We are no longer so much concerned about increasing production, but about controlling reproduction; our major political preoccupations are how to regulate the spaces between bodies, to monitor the interfaces between bodies, societies and cultures, to legislate on the tensions between habitus and body. We want to close up bodies by promoting safe sex, sex education, free condoms and clean needles. We are concerned about whether the human population of the world can survive global pollution. The somatic society is thus crucially, perhaps critically, structured around regulating bodies.

REFERENCES

Abercrombie, N., S. Hill and B.S. Turner (1980) *The Dominant Ideology Thesis*, London: Allen & Unwin.

Baudrillard, J. (1988) *America*, London: Verso.

Berger, P.L. (1969) *The Social Reality of Religion*, London: Faber and Faber.

Bogner, A. (1989) *Zivilisation und Rationalisierung, die zivilisationstheorien M. Webers, N. Elias und der Frankfurter Schule*, Opladen: Westdeutscher Verlag.

Elias, N. (1978) *The Civilizing Process: the History of Manners*, Oxford: Basil Blackwell.

Elias, N. (1991) *The Society of Individuals*, Oxford: Basil Blackwell.

Featherstone, M. (1991) *Consumer Culture and Postmodernism*, London: Sage.

Featherstone, M., M. Hepworth and B.S. Turner (eds) (1991) *The Body, Social Process and Cultural Theory*, London: Sage.

Foucault, M. (1979) *Discipline and Punish, the Birth of the Prison*, Harmondsworth: Penguin.

Foucault, M. (1981) *The History of Sexuality; Volume One: an Introduction*, Harmondsworth: Penguin.

Foucault, M. (1985) 'La vie: l'experience et la science', *Revue de metaphysique et de morale*, vol. 90:3–14.

Giddens, A. (1984) *The Constitution of Society. Outline of the Theory of Structuration*, Cambridge: Polity Press.

Giddens, A. (1991) *Modernity and Self-Identity. Self and Society in the Late Modern Age*, Cambridge: Polity Press.

Gould, M. (1991) '*The Structure of Social Action*, at least sixty years ahead of its time', pp. 85–107 in R. Robertson and B.S. Turner (eds) *Talcott Parsons, Theorist of Modernity*, London: Sage.

Heller, A. (1984) *Everyday Life*, London: Routledge & Kegan Paul.

Hennis, W. (1988) *Max Weber, Essays in Reconstruction*, London: Allen & Unwin.

Hepworth, M. and B.S. Turner (1982) *Confession, Studies in Deviance and Religion*, London: Routledge & Kegan Paul.

Honneth, A. and H. Joas (1988) *Social Action and Human Nature*, Cambridge: Cambridge University Press.

Kroker, A. and M. Kroker (eds) (1987) *Body Invaders, Panic Sex in America*, Montreal: New World Perspectives.

Marx, K. (1970) *Capital*, London: Lawrence & Wishart, 3 vols.

Sica, A. (1988) *Weber, Irrationality and Social Order*, Berkeley: University of California Press.

Sohn-Rethel, A. (1970) *Geistige und Körperliche Arbeit. Zur Theorie der gesellschaftlichen Synthesis*, Frankfurt: Suhrkamp.

Stauth, G. and B.S. Turner (1988) *Nietzsche's Dance, Resentment, Reciprocity and Resistance in Social Life*, Oxford: Basil Blackwell.

Turner, B.S. (1981) *For Weber, Essays on the Sociology of Fate*, London: Routledge & Kegan Paul.

Turner, B.S. (1982) 'The government of the body; medical regimens and the rationalisation of diet', *British Journal of Sociology*, vol. 33:367–91.

Turner, B.S. (1983) *Religion and Social Theory, a Materialist Perspective*, London: Heinemann Educational Books.

Turner, B.S. (1984) *The Body and Society, Explorations in Social Theory*, Oxford: Basil Blackwell.

Turner, B.S. (1987) *Medical Power and Social Knowledge*, London: Sage.

Turner, B.S. (1992) *Max Weber, From History to Modernity*, London: Harper Collins and Routledge.

Weber, A. (1920–1) 'Prinzipielles zur Kultursoziologie: Gesellschaftsprozess, Zivilisationprozess und Kulturbewegung', *Archiv fur Sozialwissenschaft und Sozialpolitik*, vol. 47:1–49.

Wolpert, L. (ed.) (1991) *Biological Asymmetry and Handedness*, Chichester: John Wiley & Sons.

Wrong, D.H. (1961) 'The oversocialized concept of man in modern sociology', *American Sociological Review*, vol. 26:183–93.

Introduction

WHAT IS A BODY?

One way of reading these essays is to see them as an exploration of the historical and social consequences of the management of the body in human affairs. A fundamental premise of this sociology is that the body represents a regulatory problem in the development of human civilizations. For a variety of reasons relating to the unfinished nature of human beings at birth, which have been explored by philosophical anthropologists such as Arnold Gehlen, human bodies have to be trained, manipulated, cajoled, coaxed, organized and in general disciplined. The training or cultivation of bodies by disciplines is a principal feature of culture as learned behaviour. There is a variable relationship in human behaviour between the trained and the untrained, a variation which can ultimately only be the topic of empirical research. To take one example, Norbert Elias (1991) has pointed out a number of sociologically interesting features of the smile. While young babies have an innate capacity to smile, the smile is far more malleable in adults, because we need to know the appropriate conditions for smiling and how to differentiate between a friendly smile and a sardonic leer. However, we still retain an almost irrepressible inclination to return a smile to a smiling face. This simple illustration pinpoints the idea that social life depends upon the successful presenting, monitoring and interpreting of bodies. Handshakes, winks, salutes, attention, bending and walking correctly are, in the words of Marcel Mauss (1979), 'body techniques'. Mauss showed that common bodily activities such as walking require an organic foundation, but they are also socially learned and culturally variable across societies.

By arguing that there are organic foundations to this problem of regulation, I am not adopting a position of naive materialist causality. My position, which in part I derived from Oliver Sacks in my *The Body and Society* (Turner 1984), is that we can conceive of the body as a potentiality which is elaborated by culture and developed in social relations. Thus, we do not have to pose an absolute dichotomy between acquired and innate behaviour, between culture and nature:

> Walking, at its most elementary, is a spinal reflex, but it is elaborated at higher and higher levels until, finally, we can recognize a man by the way he walks, by *his* walk.
>
> (Sacks 1981:224)

Throughout this volume, this problem of the appropriate conceptualization of the body as organism, the body as potentiality, the body as a system of representation and the body as a lived experience will constantly recur. In various chapters, I propose a number of interrelated solutions to this problem, which I call at various points epistemological pragmatism, the question of levels of analysis and the theoretical strategy of inclusion. Most of these issues are explored in the interview with Richard Fardon, where the merits and difficulties of these solutions are debated. At this stage, I am merely indicating the direction taken by this volume, which is an attempt to avoid the exclusionary division between foundationalist and anti-foundationalist ontologies, and constructionist and anti-constructionist epistemologies. In brief, it appears to be bizarre to argue that there are no organic foundations to human activity. For example, it is unlikely that a human being will ever outrun a horse over a mile under fair conditions; if the front legs of the horse are not tied together! Although it has been argued that Apache Indians were a formidable enemy of the American cavalry because they could often outdistance a horse over a three-day march, the Apaches did not have to carry their own feed. Despite these examples of human prowess, the body is a limit. However, it is wrong to think of the body as *simply* part of the environment, because human activities are embodied, as I try to argue in the chapter on 'The Absent Body in Structuration Theory'. At various stages in writing about the body, I have adopted an idea, which was originally developed by Peter L. Berger from the work of Gehlen, that we can talk about having a body, being a body and doing a body. In German, part of this distinction is present in the contrast between the lived, experiential body and the instrumental body, that is between *Leib* and *Körper*.

It is equally the case that the body is used to represent social phenomena. The idea of a corporation is a metaphor which originally depended for its force on an analogy between body and social functions. The body is simultaneously, conjointly and concurrently socially constructed and organically founded. The fact that the human body is fabricated should come as no surprise to sociology, but look at this word 'fabric'. Medieval writers typically used architectural metaphors to describe the structure of the body, while the membrane was described in Old French as *pannicle* (*panniculs*) from the word *pan* for a piece of cloth. Thus the membrane was seen as the cloth of a body (Pouchelle 1990). *Skeletos* (dried up) provides us with skeleton, but these metaphors can travel in many directions. While architecture was used as metaphors for the skeleton, we can also talk about a skeletal design for a building. Similarly we may talk about the wing of a building. The metaphors of the body have in particular provided a rich vocabulary for political and social thought over the centuries – a topic much explored by philosophers, historians and anthropologists, and sociologists (Laqueur 1990; O'Neill 1985; Turner 1991). This metaphor of 'the government of the body' is the topic of Chapter 6.

SACRED AND PROFANE BODIES

My approach to the body has been as far as possible historical and comparative, and my original interest in the body as a topic of research grew out of an earlier involvement in the sociology of religion, which produced two books in which religion and the body emerged as critical issues. In the first with Mike Hepworth, I examined the sociology of confession in deviance and religion (Hepworth and Turner 1982). We treated confession as part of a broader apparatus of moral control in society by which people are included and excluded. Since in medieval times many of our specialized distinctions between deviance, sickness, evil and unlawfulness simply did not exist, the rituals of inclusion and exclusion were a relatively undifferentiated mechanism of social policing. The treatment of lepers is a case in point. Leprosy was regulated by an ensemble of religious, legal and social norms, which prohibited the inheritance of property, marriage, residence and so forth. However, like many terrifying diseases, leprosy could be

regarded as either a punishment for sin or as a mark of divine intervention; it was both an offence and a sacred malady (Hepworth and
Turner 1982:27). In Chapter 7 of this volume, I consider the anatomy
lesson as simultaneously a moral, medical and juridical practice.

The body has been a target of many diverse social practices,
which aim at regulating the body. In *Religion and Social Theory*
(Turner 1983), where I was especially influenced by Max Weber's
discussion of 'theodicy' in relation to human suffering, sexuality
and death, I attempted to examine these regulative practices within
the broad historical context of secularization. The concept of *religion*
is related to the ideas of bonding and ruling; religion from a
Durkheimian perspective binds us into social groups and from a
Weberian perspective it authoritatively regulates social relations.
The body is also bound into society and regulated by culture. It
is important not to accept a static view of this relationship, and
any analysis of religion must take up the problem of secularization.

In order to avoid a simplistic view of secularization as religious
decline, it can be argued that many of the regulative moral functions of religion have been transferred to medicine, which polices
social deviance through the creation of a sick role in the doctor-
patient relationship. Some of these ideas were explored in *Talcott
Parsons on Economy and Society* (Holton and Turner 1986:109–42)
and in *Medical Power and Social Knowledge* (Turner 1987a:18–38).
The growing importance of preventive medicine and the use of the
concept of 'life-style' to regulate employees in order to manage
corporate insurance demands have meant that there is a major
intervention of medical ideas and practice into everyday reality –
through diet, exercise, anti-smoking norms, sexual regulation of
appropriate (that is 'healthy') partners, the regulation of childbirth,
and the hygienic treatment of death. In these areas of family life,
the local GP has replaced, in functional terms, the confessor and
priest.

Of course, this medical management has a differential impact
depending on gender, class and age. In my work on the sociology
of the body, I have been particularly interested in the idea of the
regulation of female sexuality, and hence the patriarchal management of female bodies by church and state. One aspect of this
regulation historically was the connection between economic
production and sexual reproduction. Feudalism and early capitalism
required the regulation of women in inheritance systems which were
based on primogeniture; economic success depended on the

production of a line of legitimate male heirs. Hence, virginity at marriage and fidelity within wedlock were conditions of economic prosperity (Abercrombie *et al.* 1980:59–94). To this economic imperative was added the legacy of sexual teaching from both the classical world and Christianity (Rousselle 1988) in which women's bodies are a source of pollution. The regulation of female bodies became an important part of the 'training' of saints in the medieval Catholic Church, and the self-regulation of the body a feature of women's opposition to ecclesiastical regulation (Brumberg 1988). Some features of this paradox are considered in Chapter 8 on the problematic nature of anorexia.

Although much of the theoretical framework for this interest in the historical regulation of female sexuality was derived from Weber's analysis of the processes of rationalization, this approach is also significantly influenced by the work of Michel Foucault on the body, discipline and surveillance. There are striking parallels between Weber and Foucault. Both were significantly influenced by the philosophy of Nietzsche (Stauth and Turner 1988), and both men were concerned to understand the nature of rationality and truth in relation to different belief systems. Although they used a different vocabulary, Weber's interest in how 'personality' is produced has a relationship to Foucault's ideas about the 'techniques of the self' (Foucault 1988a). Both men were engaged in an analysis of how army and religious discipline produced a regulation of the body and the self (Turner 1985 and 1987b).

This interpretation of the relationship between Weber and Foucault, and the more general thesis about medicalization as rationalization, have been challenged by a number of critics. Malcolm Bull (1990) claims that I have misrepresented the relationship between Weber and Foucault, and that I have not provided as yet a clear account of the connections between rationalization, secularization and medicalization. It is claimed that Foucault's argument is that it is discourse which produces the body as problem and not the problem of the body which produces the discourse. In short, Foucault does not hold a foundationalist view whereas Weber does. Bull's argument is important, but I can only deal with his critique in a superficial manner. I consider his objections in sequence.

Interpretations of Foucault's work are obviously problematic, for reasons which are not accidental to the nature of Foucault's enterprise. Because Foucault himself encouraged open readings of texts, it would be odd to insist upon only one orthodox reading of Foucault's work. My own view is that Foucault's work requires a non-discursive Other which stands beyond the representations of systems of knowledge, which provides as it were a platform for ideas about resistance, opposition and desire (Boyne 1990). The body is part of that resistance to the system. Although Foucault does argue that the body is the effect of discourses and disciplines, he appears to have another version of the body which is part of his view of Otherness. In addition, at a less abstract level, there are aspects of Foucault's philosophy which are compatible with my use of Foucauldian ideas and which do make Foucault's work read like a version of Weber's analysis of rationalization. In particular, I have on a number of occasions made use of Foucault's discussion of 'The politics of health in the eighteenth century' in the collection edited by Colin Gordon, where he takes a definite view of the historical consequences of demographic growth:

> The great eighteenth-century demographic upswing in Western Europe, the necessity of co-ordinating and integrating it into the apparatus of production and the urgency of controlling it with finer and more adequate power mechanisms cause 'population' with its numerical variables of space and chronology, longevity and health, to emerge not only as a problem but as an object of surveillance, analysis, intervention, modification, etc.
> (Foucault 1980:171)

Within this framework of surveillance, the body of individuals and the body of populations arise in social science as a consequence of political problems of urban management. It was on the basis of this distinction that I identified four societal problems – reproduction, restraint, regulation and representation – in *The Body and Society* (1984:91). Although Foucault often refers to 'the body' in quotation marks to distance himself, I assume, from any simplistic understanding of the human organism within a biological or natural science framework, Foucault does occupy himself with powerful descriptions of the transformations of actual bodies in time and space – for example the descriptions of judicial torture in *Discipline and Punish*, the murders which occur in *I Pierre Riviere*, or the pathological bodies in *The Birth of the Clinic*. Now, in these cases, Foucault wants to show how 'bodies' are constituted by discipline, by bureaucratic decisions and by the medical gaze, and yet the terrifying descriptions of bodies and the sufferings also have an effect which is to indicate the Otherness of life.

In reading Foucault, it is difficult not to notice a moral disgust at the consequences of rationalization and secularization in destroying the presence of difference by making the world the same. One can read Foucault as a critic of normalization, especially the construction of the 'normal' in bureaucratic practice. This issue in Foucault is, from my perspective, another version of the contrast in Nietzsche between Apollo and Dionysus, and the opposition in Weber between charisma and rationalization. In my interpretation, therefore, Bull is wrong to object that secularization 'is not a form of rationalization' (Bull 1990:248), but his objection does offer me the occasion to spell out more clearly the ideas of secularization and medicalization as rationalization. I have consistently held the view that Weber's sociology is held together by a philosophical anthropology (Turner 1981). Capitalist rationalization is not a 'natural' state of affairs, because it forces human beings to behave in ways which are not necessary for survival, that is it forces people to produce more than they need. In *The Protestant Ethic and the Spirit of Capitalism*, Weber (1930:60) argues

> A man does not 'by nature' wish to earn more and more money, but simply to live as he is accustomed to live and to earn as much as is necessary for that purpose. Wherever modern capitalism has begun its work of increasing productivity of human labour by increasing its intensity, it has encountered the immensely stubborn resistance of this leading trait of pre-capitalistic labour.

The rationalization of labour, and hence of the body, has been an essential feature of the development of capitalism. This rationalization of labour in capitalism has been achieved by practices of discipline, diet, training and regulation. The whole framework may be termed, following Foucault, panopticism. I have on several occasions attempted to provide some illustrations of this 'government of the body', in particular in the work I have done on dietary regimes, for example in Chapter 6. Now one of the crucial features of the rationalization of the productive body has been medicalization, that is the rational application of medical knowledge and practice to the production of healthy, reliable, effective and efficient bodies. Much of the history of the early twentieth century was the application of scientific medicine, especially in Germany, to the production of a healthy, fit and fighting national body (Weindling 1989).

However, I also want to argue that this medicalization of the body as rationalization is a secularization of culture. Early Christianity did not make an unbridgeable separation between the salvation of the soul and the body; to salve the body was a necessary activity of grace alongside the salvation of the soul. In the classical world there was always a close relationship between medicine and ethics. As Foucault (1988b:55–6) observed in his study of 'the cultivation of the self' in writers such as Plutarch, Marcus Aurelius and Seneca:

> Educating oneself and taking care of oneself are interconnected activities. . . . The increased medical involvement in the cultivation of the self appears to have been expressed through a particular and intense form of attention to the body.

The development of society in terms of the differentiation of religion, medicine, law and government eventually brought about a separation of the functions of the doctor and the priest, and then a transfer of moral regulation from the church to the clinic. This transfer does not mean that medicine serves some non-rational purpose; it means that the confessional as a moral regulation (primarily of women) has become more 'scientific' through the development of Freudianism and clinical psychiatry, but the effect of these new practices is similar to that of clinical theology and pastoral theology. Medical practice in our time clearly does have a moral function, especially in response to AIDS and IVF programmes for unmarried, single women, but these moral functions are typically disguised and they are ultimately legitimized by an appeal to scientific rather than religious authority. Diet and jogging are seen to contribute to health and longevity; they are not overtly recommended as ascetic practices which are beneficial to the soul. Thus, in Foucauldian terminology, medicine occupies the social space left by the erosion of religion. This argument is obviously controversial and one issue which would be worth considering is that, especially in America, organized religion has itself appropriated the rational techniques of medicine, commercial advertising, psychiatry and clinical sociology to produce a 'religious package' which is designed to meet the needs of the new middle class. For young Texan women, there may be a relationship between a 'high' in strenuous exercise and a religious 'high'.

THE BODY AND SOCIAL THEORY

This volume is, therefore, concerned to advance a number of substantive arguments about the historical development of the government of the body, but I also want to claim that this emphasis on the body and embodiment has important implications for the development of sociology, and in particular medical sociology. These prescriptive features of this programme are advanced in Chapters 4 and 5. These essays have to be seen as part of a more general attack on Cartesian dualism in the sociology of social action, in which the principal emphasis is given to cognitive activity, especially the selection of means which are appropriate to known goals. There appears to be a rationalistic bias in sociology which has until recently treated the social actor as a disembodied rational agent. In these conventional approaches to the nature of social action from Weber onwards, the body appears in sociological theory as a feature of the environment. There are some major exceptions to this argument in the work, for example, of Erving Goffman and G.H. Mead. More recently, Pierre Bourdieu has made an important contribution to the emergence of a sociology of the body, and I attempt to examine his concepts of habitus and practice in relation to the symbolic power of the body in Chapter 2. In general, I have been critical of Giddens's neglect of embodiment in his development of a theory of structuration. However, in his recent book on *Modernity and Self-Identity*, there is a brief commentary on the nature of the body, which he suggests is no longer an extrinsically 'given' in modernity; it is itself 'reflexively mobilized' (Giddens 1991:7). I am not convinced, however, that these comments on the body are sufficient to outweigh the predominance of 'praxicology' in Giddens's structuration theory. The core of the theory is the reflexive self not the embodied self. It is also not clear whether this reflexively mobilized body is specific to modernity. We have already seen how Foucault analysed the reflexive importance of the body in Greek medical ethics and in Stoical philosophers from Roman times. We could also mention the much neglected figure of Benjamin Nelson who traced the origins of this body–self reflexivity to the origins of the confessional self in medieval religious practices, such as confession (Nelson 1981).

Turning from sociology in general to the specific area of medical sociology (or more accurately the sociology of health and illness), I have argued in Chapter 5 that a sociology of the body is a fundamental basis for the *theoretical* of medical sociology. Although medical sociology has been, in institutional terms, one of the most successful fields of sociology, it has become increasingly separated from mainstream sociology. In the past sociologists like Parsons showed how medical institutions and practices were of central theoretical concern to the whole of sociology. The same argument could be sustained with respect to the work of sociologists like Anselm Strauss, Barney Glaser and Julius Roth, whose contributions to the medical field (in research on death and time) were also seen to be highly relevant to mainstream or core issues for sociology as a whole. The leading figures of medical sociology today appear to make few contributions to the development of sociology as such. The nature of funding for medical sociology and the location of medical sociologists in medical schools often have the negative consequence of creating a division between sociological theory and the practice of medical sociology. In this volume, I attempt to elaborate arguments which were originally developed in *Medical Power and Social Knowledge* that one solution to this hiatus would be to develop a sociology of the body as a theoretical programme in medical sociology as a bridge with theoretical sociology.

In thinking about the origins and development of sociology, we have often neglected its relationship to the practice of medicine. This neglect is probably a function of the critical relationship between the sociology of health and illness and Flexnerian medicine, that is twentieth-century medicine, which is organized around a collection of basic natural sciences in terms of 'the medical model'. This model assumes that illness can be explained in terms of determinate causes operating on the body, which is conceptualized as a machine. In Chapter 4, I argue that the development of Flexnerian medicine was at the cost of social medicine and that sociology clearly has historically and analytically an important relationship to social medicine. Sociologists are typically critical of the medical model, because it is not directly interested in the social context of disease and its orientation is to treat the body of the patient as an object of scientific inquiry and intervention. Furthermore, Flexnerian clinical practice treats the body as an ensemble of specialized parts, which require separate specialized interventions. The medical model negates the idea of the patient as an embodied subjectivity. There are good reasons, therefore, for sociology to take seriously the ideas of phenomenology in perceiving the body as a 'lived body' in order to develop simultaneously a critique of rationalistic social action theory and the medical model.

CONCLUSION

Feminist social theory and postmodernism have in recent years had a profound impact on social theory. Feminism has been especially important in making problematic the relationship between gender and sex, between culture and nature. These social divisions, which have been topics of feminist theory, are now rendered increasingly problematic by the development of cyborgs (cybernetic organisms), which involve an integration of living organisms and robotic devices. The political and cultural issues which are generated by these technological developments have been analysed by, for example, Donna J. Haraway (1990) in *Simians, Cyborgs, and Women*.

The critique of Cartesian dualism, the postmodernization of our conceptual apparatus, the deconstruction of 'grand narratives' and feminist debates about the inadequacy of conventional (male) views of embodiment have, in recent years, converged around the idea of difference. Against the modernist idea of universal history and unitary experiences, postmodernism has celebrated pluralism, difference, play and parody (Turner 1990). In the sociology of the body, this critical movement has primarily produced a consensus view that the body is socially constructed. For example, the representation of sexual organs in medical discourse reflects dominant conceptions of the role of the (two) sexes in society (Laqueur 1990). Our language for describing reproduction ('labour') has been shaped by the dominant ethic of industrial capitalism (Martin 1989). The location of diseases within social space and the space of the body is an effect of changes in the medical gaze (Foucault 1973).

These approaches are now familiar in social theory and some aspects of the debate have influenced the work of medical sociologists. Although I find these theoretical developments both exciting and important, I have also attempted to retain in this volume a commitment to the legacy of phenomenology and philosophical anthropology as an interpretation of the 'lived body'. The sociological understanding of the experience of illness has an important part to play in the development of the sociology of health and illness, but it also contributes to moral debate (Frank 1991). The relationship between constructionism and foundationalism is discussed at a number of points in this volume, especially in

Chapters 1, 2, 3 and 9. I shall not repeat those arguments in this introduction. However, it is the case that many of these 'new' discussions are in fact merely versions or repeats of previous debates in the social sciences. For example, as a consequence of the impact of ethnomethodology on sociology in the 1960s and 1970s, it became fashionable to argue that sentences (or truth claims) only made sense in their context. Sentences ('The cat sat on the mat') have a function which cannot be understood by reference to objectively existing phenomena ('There is a cat!'), but only by reference to their place in a conversation or text (about 'cats' and 'mats'). In short, statements were held to be indexical.

One important criticism of the doctrine of indexicality was that some statements might be more indexical than others. That is, statements like '2 + 2 = 4' may be less indexical than 'Young thin girls may be anorexic'. In common-sense terms, some statements are more context-dependent than others, because their meaning is deeply embedded in cultural fields which in addition may be charged with considerable ideological presuppositions. Ideas like 'young' or 'thin' or 'anorexic' are culturally ambiguous and problematic; they may not be open to translation. They function within a political discourse of morality and medicine. In this volume, I adopt a rather similar position about constructionism. I see no reason to deny the idea that medical or psychiatric concepts such as anorexia or hysteria or anxiety are socially constructed. I see no reason to doubt the proposition that the body is socially constructed. However, some things ('hysteria') may be more socially constructed than others ('gout'). Secondly, topics which are politically charged (such as 'black lung') are more likely to be regarded as socially constructed by sociologists than other conditions (goitre). Thirdly, it is obviously a mistake to assume that different approaches to the body are talking about the same 'thing'. Feminist critique of the representations of the body is not addressing the same issue as, for example, Merleau-Ponty's phenomenological analysis of the experience of phantom-limb pain. The assumption of this book is that a phenomenology of the body can have a practical importance for medical sociology as an applied subject, because it provides a sophisticated but sensitive perspective on issues like pain, disability or death. Finally, there is a moral debate implicit in this volume, which concerns the frailty of the human body in relation to self-identity and social significance.

REFERENCES

Abercrombie, N. (1975) 'Sociological indexicality', *Journal for the Theory of Social Behaviour*, vol. 4(1):89–95.

Abercrombie, N., S. Hill and B.S. Turner (1980) *The Dominant Ideology Thesis*, London: Allen & Unwin.

Boyne, R. (1990) *Foucault and Derrida. The Other Side of Reason*, London: Unwin Hyman.

Brumberg, J.J. (1988) *Fasting Girls, the Emergence of Anorexia Nervosa as a Modern Disease*, Cambridge, Mass.: Harvard University Press.

Bull, M. (1990) 'Secularization and medicalization', *British Journal of Sociology*, vol. 41(2):245–61.

Elias, N. (1991) 'On human beings and their emotions: a process-sociological essay', pp. 103–25 in M. Featherstone, M. Hepworth and B.S. Turner (eds) *The Body, Social Process and Cultural Theory*, London: Sage.

Foucault, M. (1973) *The Birth of the Clinic, an Archaeology of Medical Perception*, London: Tavistock.

Foucault, M. (1980) 'The politics of health in the eighteenth century', pp. 166–82 in C. Gordon (ed.) *Power/Knowledge, Selected Interviews and Other Writings 1972–1977*, Brighton: Harvester Press.

Foucault, M. (1988a) 'Technologies of the self', pp.16–49 in L.H. Martin, H. Gutman and P.H. Hutton (eds) *Technologies of the Self, a Seminar with Michel Foucault*, London: Tavistock.

Foucault, M. (1988b) *The History of Sexuality, the Care of the Self*, Harmondsworth: Penguin Books.

Frank, A. (1991) *At the Will of the Body, Reflections on Illness*, Boston: Houghton Mifflin.

Giddens, A. (1991) *Modernity and Self-Identity. Self and Society in the Late Modern Age*, Cambridge: Polity Press.

Haraway, D.J. (1990) *Simians, Cyborgs, and Women: the Reinvention of Nature*, London: Free Association Books.

Hepworth, M. and B.S. Turner (1982) *Confession, Studies in Deviance and Religion*, London: Routledge & Kegan Paul.

Holton, R.J. and B.S. Turner (1986) *Talcott Parsons on Economy and Society*, London: Routledge & Kegan Paul.

Laqueur, W. (1990) *Making Sex, Body and Gender from the Greeks to Freud*, Cambridge, Mass. and London: Harvard University Press.

Martin, E. (1989) *The Woman in the Body, a Cultural Analysis of Reproduction*, Milton Keynes: The Open University Press.

Mauss, M. (1979) *Sociology and Psychology: Essays*, London: Routledge & Kegan Paul.

Nelson, B. (1981) (edited by T.E. Huff) *On The Roads to Modernity. Conscience, Science and Civilizations. Selected Writings by Benjamin Nelson*, Totowa, New Jersey: Rowman and Littlefield.

O'Neill, J. (1985) *Five Bodies, the Human Shape of Modern Society*, Ithaca and London: Cornell University Press.

Pouchelle, M-C. (1990) *The Body and Surgery in the Middle Ages*, New Brunswick, New Jersey: Rutgers University Press.

Rousselle, A. (1988) *Porneia, on Desire and the Body in Antiquity*, Oxford: Basil Blackwell.

Sacks, O. (1981) *Migraine, Evolution of a Common Disorder*, London and Sydney: Pan Books.

Stauth, G. and B.S. Turner (1988) *Nietzsche's Dance, Resentment, Reciprocity and Resistance in Social Life*, Oxford: Basil Blackwell.

Turner, B.S. (1981) *For Weber, Essays on the Sociology of Fate*, London: Routledge & Kegan Paul.

Turner, B.S. (1983) *Religion and Social Theory, a Materialist Perspective*, London: Heinemann Educational Books.

Turner, B.S. (1984) *The Body and Society, Explorations in Social Theory*, Oxford: Basil Blackwell.

Turner, B.S. (1985) 'The practices of rationality: Michel Foucault, medical history and sociological theory', pp. 193–213 in R. Fardon (ed.) *Power and Knowledge, Anthropological and Sociological Approaches*, Edinburgh: Scottish Academic Press.

Turner, B.S. (1987a) *Medical Power and Social Knowledge*, London: Sage.

Turner, B.S. (1987b) 'The rationalization of the body: reflections on modernity and discipline', pp 222–41 in S. Whimster and S. Lash (eds) *Max Weber, Rationality and Modernity*, London: Allen & Unwin.

Turner, B.S. (ed.) (1990) *Theories of Modernity and Postmodernity*, London: Sage.

Turner, B.S. (1991) 'Missing bodies – towards a sociology of embodiment', *Sociology of Health & Illness*, vol. 13(2): 265–72.

Weber, M. (1930) *The Protestant Ethic and the Spirit of Capitalism*, London: Unwin University Books.

Weindling, P. (1989) *Health, Race and German Politics between National Unification and Nazism 1870–1945*, Cambridge: Cambridge University Press.

Part I

Discovering bodies

The body question
Recent developments in social theory

AN INTRODUCTION TO THE SOCIOLOGY OF THE BODY

In the past decade, both the social sciences and the humanities have turned increasingly to an exploration of the problem of the body in social life in order to understand the complexity of our particular historical conjuncture. In this respect, the work of Michel Foucault and the revival of interest in Nietzsche have been important intellectual developments. Although there are many novel elements to this debate, the issue of the body in human societies has in fact been a persistent theme of Christian culture in the West. The apparently simple questions (What is the body? What is embodiment?) have persistently and perennially dominated academic and public discussion for reasons which are considered in this introductory comment, and throughout this volume. My principal aim here is to offer an introduction to this debate about the body, and to suggest various reasons why this topic is of crucial importance as a focus of research in the social sciences, but in more specific terms I want to suggest that a sociology of the body is an essential underpinning for the sociology of religion and medical sociology. These sub-fields engage with issues, namely theodicy and human suffering, in which the frailty of the human subject as a consequence of embodiment is an unavoidable issue. In fact, frailty is probably the most promising theme for a minimal theory of ontology from a sociological perspective.

Before getting into this debate, it is necessary to consider the absence of the body from traditional sociology, and to criticize many of the existing assumptions about the relationship between mind and body, which have dominated both the medical and the social sciences since at least the seventeenth century. For some philosophers, this subordination of the body under an ethic of world mastery, of which science and technology are major components, is one of the defining characteristics of Western civilization as such, and as a result this problematic status of the body/nature is part of the post-Socratic world of rationalism. This problem of embodied being in relation to technology and rationality can be regarded as *the* problem of Western philosophy (Heidegger 1989). It was on this basis that Nietzsche in *The Will to Power* rejected the 'soul-hypothesis' and proposed to start (philosophy) again from the premise of the body (Stauth and Turner 1988). This study of the body in the realms of medicine, politics and religion is consequently based on the assumption that the traditional mind/body dichotomy and the neglect of human embodiment are major theoretical and practical problems in the social sciences.

The social sciences have in general accepted the Cartesian legacy in which there is sharp division between the body and mind. Cartesian dualism is based on the principal assumption that there is no interaction, or at least no significant interaction, between mind and body, and therefore that these two realms or topics can be addressed by separate and distinctive disciplines. The body became the subject of the natural sciences including medicine, whereas the mind or *Geist* was the topic of the humanities, or the cultural sciences (*Geisteswissenschaften*). This separation was consequently an important feature of the very foundation of the social sciences, especially in the debate about the relevance of (natural) science methodology for the interpretative sciences of 'Man'. This problem exercised Max Weber more or less continuously in the development of his epistemology in the debate about an appropriate *Wissenschaftslehre* (Weber 1949).

It was this dualism in the Western conceptualization of science which eventually legitimized various forms of reductionism in which mental events, the life of the spirit and culture are explained by, or in terms of, material causes. It is also common to hear in everyday language and in health discourse that 'something' (anorexia, repetitive strain injury, miner's lung or agoraphobia) does not 'exist' because it is 'only in the mind'. The concept of 'psychosomatic illness' does not help in these circumstances, because in common parlance it still means 'only in the mind'; the expression still preserves a mind (psycho)/body (somatic) duality. Thus, the Cartesian division in the medical sciences allowed medicine to treat the problems of the body with minimal reference to social or psychological causes, especially after the curriculum reforms that followed the Flexner Report in 1910. This duality also provided the legitimation for regarding the apparent success of alternative medical world-views such as acupuncture or homoeopathy as a consequence *merely* of the placebo effect. As a result, allopathic medicine has in general shown little interest in what philosophers refer to as 'the lived body' as opposed to the objective body (Leder 1990).

Although Cartesianism had these characteristics (dualism, reductionism and positivism), it is perhaps ironic that contemporary interpretations of the philosophy of Descartes, especially *The Discourse on Method*, have claimed that Descartes's own position was characterized as 'dualistic interactionism' (Wilson 1978). It is clear from a close reading of *The Discourse on Method* that Descartes believed that there was in fact a close interaction between the body and the mind, and disease was the consequence of a disturbance in this interaction; the role of medicine was to resolve the problems in this interdependency (T. Brown 1985). However, Descartes's 'dualistic interactionism' eventually evolved in the natural sciences into a unitary and positivistic view of materialism in which the disciplines which attempt to develop explanations of events in nature and society, body and mind, environment and culture were both isolated and specialized.

Despite Descartes's own version of interactionism, the consequences of the Cartesian legacy have been very significant for both the natural and the social sciences. In this introduction I shall focus mainly on sociology, where the notion of the social actor and social action have been primarily and classically developed within this dualistic Cartesian framework. By treating the body as part of the environment of action, sociology was developed as an interpretative science of the meaning of action in the methodology of Weber; sociology was a discipline within the *Geisteswissenschaften* whose aim was the cultural understanding of the shared meaning of action.

The importance of the success of economics as a science of rational (economizing) action in shaping the early development of sociology, especially in the work of Weber and Pareto, has often been neglected in the history of the discipline. This critical interaction between sociology and economics was particularly formative in the work of Talcott Parsons (Holton and Turner 1986). Sociology was driven in part by attempts to understand the role of values and knowledge in economic choices. In its emphasis on voluntarism, choice and action (Parsons 1937), sociology placed a special importance on the idea of consciousness or knowledge-ability of the social actor (Giddens 1984). The principal defining characteristics of *homo sociologus* were, first, the importance of shared meanings to define a social situation and, second, the presence of knowledge and understanding whereby the social actor has an awareness of means and ends. The knowledgeable actor is one who selects between different goals in terms of values and appropriate means in terms of norms. This combination of dimensions was developed classically by Parsons as a critique of economics (Robert-son and Turner 1989). Anthony Giddens's 'structuration theory' (Giddens 1984) is in many respects very different from Parsons's 'voluntaristic theory of action', because, where Parsons was con-cerned to understand how values are shared (by the processes of internalization and socialization), Giddens has been concerned to understand human action in terms of its reflexivity. Human action is primarily self-monitoring action; human beings cannot avoid the constant confrontation of choice. Consequently, neither Parsons nor Giddens has shown much concern for the embodiment of the human actor. In Parsons's sociology of action, the body is part of the environment of action in his analysis of the unit act and the social system; in structuration theory, Giddens, following the theories of the geographer T. Hagerstrand, treats the body as an aspect of the time–space constraints on human action (Urry 1991).

As a consequence of this interest in the rational and non-rational nature of social action, sociological theory has effectively neglected the importance of the human body in undersanding social action, and social interaction. The nature of human embodiment has, with some important exceptions, not been important in either social research or social theory. As a consequence, the body has been curiously missing or absent from sociological thought (Turner 1991a). Until recently, this absence was true of such sub-disciplines as the sociology of health and illness, where one might imagine in common-sense terms that a discussion of health without any presuppositions about the body would have been impossible (Turner 1987). One might also imagine that in the sociology of religion, where the question of theodicy in relation to issues concerning death, disease and sexuality is an analysis of the body as 'flesh', the centrality of the body would have been a topic of major concern, but this has not been the case (Turner 1983). Within the last decade,

there has fortunately been evidence of a major interest in the
sociology of the body (Deleuze 1983; Feher 1989; Frank 1991;
O'Neill 1985; Suleiman 1986; Turner 1991b; Zola 1991), and I
hope to reflect upon this growth of interest in this particular volume.

The absence of the body from social theory is not an unimportant
or insignificant lacuna. The absent body implies and poses major
problems for the formulation of a sociological perspective on the
human agent, agency and human embodiment. If we adopt the
idea of sociology as a scientific study of action, then we require
a social theory of the body, because human agency and human
interaction involve far more than mere knowledgeability, intentionality and consciousness. Of course, this statement raises various
questions in sociology as to what is to count as an 'agent'. We need
to avoid the conventional conflation of the 'people' with the 'parts'
by being more clear about the difference between social system
analysis and social analysis (Archer 1988). If collective action refers
to social entities such as class and state, then it might be argued
that the question of the body would be relevant. However, if one
is concerned with human beings at the social rather than the system
level, then it is difficult to comprehend how sociology could avoid
the development of a sociology of the body. As a result, in this
chapter I am taking seriously Weber's claim (1978) that sociology
is the interpretive understanding of social action, and that this social
action is undertaken by embodied social beings. I also want to take
seriously the problem of the gesture in G.H. Mead's attempt to
formulate an understanding of the situated and interactive character
of the 'I' and the 'me'. For example, Mead's discussion of the
importance of the hand in gesture in relation to the central nervous
system and the origins of creative thinking is typically neglected
in subsequent accounts of the origins of a symbolic interactionist
analysis of the self. Indeed, the hand has been regarded by
philosophers like Heidegger as a crucial defining characteristic of
human beings, as an agent that shapes the environment as a
consequence of 'handwork' (Heidegger 1982:116–18).

As a consequence of their embodiment, all human agents are
subject to certain common processes which, although they have
biological, physiological and organic foundations, are necessarily
social in character. These common social processes are related to
the conception, gestation, birth, development, death and disintegration of the human body. Because many social practices and rituals
are ultimately based upon these obvious, everyday events (such

as marriage, burial and rituals of grief), it is peculiar that sociology has, in very general terms, neglected these practices as features of human embodiment. Social anthropology is probably an important exception to this rule (Turner 1991b). By concentrating on the meaningful character of social action from the standpoint of the social agent, sociology and the social sciences generally have avoided this corporeal side of human action, despite the fact that questions of meaning, such as Weber's theodicy problem (Turner 1981), are invariably associated with embodiment, that is associated with suffering, joy, death, pain and so forth.

This corporeal aspect of human agency is not in some sense beyond, alongside or outside the social. By suggesting that sociology has neglected the human body, one does not necessarily endorse any arguments in favour of biologism. The point is to avoid nineteenth-century positivism by embracing biological reductionism and to avoid idealism disguised as a theory of social constructionism. For the sociologist, the social must remain primary. Thus, in emphasizing the importance of the phenomenology of the body, it does not follow that sociology should in some way simply incorporate a biologically reductionist position. As a sociological enterprise, the sociology of the body will deal with the essentially social nature of human embodiment, with the social production of the body, with the social representation and discourse of the body, with the social history of the body, and finally with the complex interaction among body, society and culture. For reasons developed by Marcel Mauss (1979), fundamental aspects of embodied activity, such as walking, standing or sitting, are social construction. These practical activities require an organic foundation, but the elaboration of these potentialities requires a cultural context. It was for this reason that Mauss talked about 'body techniques' which, while depending upon a common organic foundation, are nevertheless both personal and cultural developments.

Perhaps more importantly for the sociology of action, the very identity of social agents cannot be easily separated from their embodiment within the interactional situation. In everyday life, in interacting with other social agents, we have in principle to be able to recognize and distinguish between different social agents. At the level of everyday life, therefore, the ongoing identification of other social agents depends fundamentally on their embodiment. In Mead's analysis of social acts and the development of the self, a 'conversation' of gestures (internal and external) was important in his understanding of the constitution of the 'I'. The face and the hand are both fundamental to such an exchange of gestures. For Mead, 'Speech and the hand go along together in the development of the social human being' (1934:237). However, it was Erving Goffman in *The Presentation of Self in Everyday Life* (1959) who showed how the representation of the disruption of order in everyday life can depend on our control over the representation of the body. The management of embarrassment can require considerable body control, if we are to avoid losing face. Incidentally, this notion of 'face' should serve to remind us how much of our social and moral language depends on the metaphors of the body: an upright person; a person of some standing; a faint-hearted soul; a person with a stiff upper lip.

Who I am rests crucially on having a specific body which I do not share with other social agents. The platitude ('I can't be in two places at once!') has major social importance. In social interaction, therefore, individuation and individuality depend upon a shared agreement about the relationship of the social actor to his or her body. Having a specific body is very important when it comes to questions of impersonation, kidnapping, paternity, legal identity, and nationality; it is for this reason that who a person is may come to depend in the last analysis on a procedure such as genetic fingerprinting. In a future society where the transplantation of organs is a routine and widespread surgical procedure, the hypothetical puzzles in classical philosophy about identities and parts will be issues of major legal and political importance. Can I be held responsible for the actions of a body which is substantially not my own body?

I have argued in general that sociology has neglected human embodiment, because it has implicitly accepted a Cartesian tradition, and because sociology has been fundamentally concerned with the social meaning of social action at the level of values and beliefs. The philosophical assumptions behind the mind/body split have been challenged by developments in philosophy which have yet to have a complete or full impact on sociology. I have tried to suggest why a proper appreciation of human embodiment is in fact an essential feature of the development of an adequate sociology of action and interaction. For example, it is difficult to talk about identity without talking about a specific body. We can individuate persons with some certainty only through fingerprints, photographs and genetics. Although memory and social records are important, to be a specific person also requires a specific body. Although this is a general problem in sociology, I believe that the absence of a coherent sociological understanding of embodiment has crucial implications for medical sociology (Turner 1987) and for a variety of other sociological fields, such as the study of human emotions, sexuality, sport, passion and ageing (Featherstone *et al.* 1991). It is in these areas (health, sport, leisure, sexuality and consumerism), where the interaction between embodiment, society and culture is a crucial feature of social practice, that we desperately need an elaborate sociology of 'the lived body'. In this particular volume on the sociology of the body, my aim is to outline the various problems which arise as a result of providing a sociological understanding of the body, to direct the reader to relevant literature in the area and to demonstrate the importance of the sociology of the body for a number of substantive fields.

PHILOSOPHICAL ANTHROPOLOGY

Although the social sciences have generally neglected the importance of human embodiment and the body, I do not want to suggest that the body has never been studied in social theory from outside the Cartesian framework. It is rather the case that a number of promising starting points for the study of the body were either abandoned or ignored. One very good illustration of this underdevelopment can be found in the (primarily German) tradition of philosophical anthropology which had its roots in the philosophy of Friedrich Nietzsche (Stauth and Turner 1988). One important dimension of Nietzsche's philosophy was the contrast which he made between Dionysus and Apollo. Dionysus was the god of sexual power, ecstasy and passion, the driving force behind the frenzied activity of the early Greek religions. Apollo was the god of order, form, rationality and coherence. Nietzsche conceived of history as the endless struggle between these two principles, but he did not argue that the resolution of this conflict had to be in the triumph of Dionysus. Nietzsche was not naively arguing for 'a return to nature'. By contrast, Nietzsche adopted the position that it was only by a successful combination of these two principles that a healthy society could emerge in which sexual passion was reconciled with the life of rational activity. In particular, Nietzsche sought this reconciliation in aesthetic activity, especially in art (Nehamas 1985). Life was to be lived as a work of art. It was partly on this

basis that Nietzsche was powerfully attracted to the music of
Wagner. Failure to produce a satisfactory synthesis of these two
principles within the life of the individual also led to disease, sickness
and madness. Nietzsche tended to think that the neuroses which
are associated with this contradiction are peculiar to human beings,
since human beings are alienated from their natural environment
by the very fact of their consciousness. As far as we know, human
beings are the only animals to reflect self-consciously on their future
death. Nietzsche regarded the human as an incomplete animal,
because human beings are not ontologically specific to any given
fixed habitat or environment, and because their completion depends
on cultural training. They change their environment to suit their
needs rather than adapting their needs to a fixed environment.
Human beings build boats rather than evolving flippers. Nietzsche
has subsequently had an enormous influence on the philosophical
development of figures as diverse as Weber, Heidegger, Freud and
Foucault (Boyne 1990; Deleuze 1983; Stauth and Turner 1988;
Turner 1981).

Nietzsche's approach to the nature of human existence was the
principal influence behind a tradition of analysis which we now
call 'philosophical anthropology' (Honneth and Joas 1988) and less
commonly 'phenomenological anthropology' (van Peursen 1956).
Within a narrow framework, this philosophical tradition included
influential writers like Arnold Gehlen (1988), Helmuth Plessner
(1976), F.J.J. Buytendijk, A. Blok, A. Portmann and J. Von
Uexkull (Honneth and Joas 1988). Within a broader framework,
the idea of a philosophical anthropology is also closely related to
the work of Max Scheler and Martin Heidegger. Indeed, any concern with the social ontology of human beings is likely to be
influenced by the legacy of Nietzsche's view of human beings as
incomplete. Thus, Heidegger's analysis of being in general was
an attempt to understand the ways in which being is always being-
in-the-world. The concreteness of the German in this respect is lost
in the English translation, because when Heidegger considered the
problem of existence or the 'being' which is appropriate to persons, the German *Dasein* (*Da-sein* or 'being there') signifies the
'placedness' and particularity of being (Heidegger 1962:27).

From a sociological point of view, the work of Gehlen and
Plessner is probably the most important and influential in
philosophical anthropology. Gehlen's study *Man* (1988), for
example, can be regarded as an extended reflection upon Nietzsche's

notion that man is a not yet completed animal or a 'not yet determined animal' (*noch nicht festegstelltes Tier*) who, as a consequence, must socially construct institutions to provide some protection and to establish some method of social continuity.[1] Gehlen thus spoke of the openness (*Weltoffen*) of human beings to their social and natural environment. Gehlen's work has been especially influential in the development of Peter L. Berger and Thomas Luckmann (1967).

Because of their ontological openness, human beings must socially construct their own reality in order to institutionalize their existence and to protect themselves against the threat of anomie or chaos. Berger has been interested throughout his sociology in the dialectical relations between the body and self, and the self and society. He has expressed this dialectical relationship in the notion that 'man' *is* a body, in the same way that this may be said of every other animal organism. On the other hand, 'man' has a body. That is, human beings experience themselves as entities that are not wholly identical with bodies but, on the contrary, have those bodies at their disposal (Berger and Luckmann 1967:48).

From a phenomenological point of view, we can make a distinction between having a body, doing a body and being a body. For example, we often experience the body as an alien environment in which our body appears as something over which we do not have control. 'It' is experienced as part of our environment. In diseased states, this experience of having a body is often prominent where the body appears as an objective and external environment (Herzlich and Pierret 1987). By contrast, we can also argue that we have, in a certain sense, a sovereign control over our own bodies in which we are embodied. In the tradition of bourgeois political philosophy in J.S. Mill, we are sovereign individuals (Abercrombie *et al.* 1986). Our relationship to our own bodies is, so to speak, so comfortable that we are not struck by its perceptual absence. In my routine daily practices of sitting, walking, sleeping and eating, I do not have to remind myself that I have a body and in normal circumstances I do not have to issue this 'lived body' with instructions such as 'Walk!' or 'Sit down!'. In this sense, I have a phenomenologically absent body (Leder 1990). Finally, following Mauss's idea of body techniques, we can think about doing the body, that is the body appears as a collection of practices over which we might have a certain mastery or sovereignty. Through childhood socialization, all of us acquire certain basic body techniques for presenting and maintaining and reproducing bodies in time and space.

I originally adopted this framework for discussing the complexity of embodiment in *The Body and Society* (1984) in which I followed much of Peter Berger's attempt to talk about the body from the point of view of the social construction of reality. However, it would be wrong to construe my sociology of the body as *merely* a social constructionist viewpoint. In arguing that the body is socially constructed (by language, ideology, discourse or knowledge), it has been assumed, wrongly in my view, that one could not in addition believe that there is such a topic as the phenomenology of pain. Disability raises problems about representation and the nature of disability results in major disputes concerning its classification. However, there is also legitimately a question as to the sociological and phenomenological reality of disability (Zola 1982). In short, I do not believe that reality is *discourse*, that is, I do not believe that social reality is merely an issue of representation.

Sociological approaches to the body must obviously be conditioned or at least influenced by the modes by which 'body' exists in the world. The question (How is the body represented in society?) is not the only question one can ask about the body. For example, we know that the sex of the body has been represented historically in many different ways, which have been determined by medical discourse (Laqueur 1990). It is however still a legitimate question to ask 'What is it like to *be* a woman?' and it is also important to understand how, for example, the absence of lactation in men influences the social process of parenting. To believe that questions of representation are the only legitimate or interesting scientific questions is to adopt a position of idealism towards the body.

In the English language, we do not have a plethora of different nouns for describing these different modes of the body. We have 'corpse' for a dead body, but no special term for the 'lived body'. The term 'embodiment' is rarely used outside an academic context. In this respect, the German language may be richer in permitting us to make important distinctions between different forms of phenomenologies of the 'lived body'. Thus, Plessner (1976) was able to contrast *der Leib* and *der Körper* as representing two dimensions of the human body. Whereas *Leib* refers to the animated living experiential body, *Körper* refers to the objective, exterior and institutionalized body. This double nature of human beings

(Plessner 1976:195) expresses the ambiguity of human embodiment as both personal and impersonal, objective and subjective, social and natural. The *Leib/Körper* distinction expresses in the language of philosophical anthropology many of the fundamental concepts in Heidegger's discussion of *Dasein* and *Sosein*. I think this is a fundamental contrast, because it precisely indicates the weakness of the Cartesian legacy in sociology, which has almost exclusively treated the human body as *Körper* rather than both simultaneously *Körper* and *Leib*. In approaching the human body as an objective and impersonal structure, sociology has by implication relegated the body to the environmental conditions of social action. *Leib* can be regarded as standing for the body-for-itself, while *Körper* is simply the body-in-itself.

The work of Gehlen was, to some extent, undermined by his association with national socialism and it is only recently that the importance of his work has been philosophically appreciated (Honneth and Joas 1988). There is also a ferocious debate around the issue of Heidegger's involvement with national socialism, especially around the problem of whether the relationship between his politics and his philosophy was contingent or necessary (Ferry and Renaut 1990; Wolin 1990). I cannot enter into that debate here; however, it is appropriate to recognize that the idea, which has been taken from Gehlen, that human beings need shelter or a 'sacred canopy' (Berger 1969) against the threat of chaos has often had very conservative implications. I have criticized this interpretation of human incompleteness in terms of Peter Berger's sociology of religion (Turner 1983).

Philosophical anthropology was part of a tradition in German sociology which was broadly referred to as *Lebensphilosophie* (or life-philosophy) and was concerned to understand the social being of humans in the world, that is, to grasp philosophically the life-world of embodied human beings. Generally speaking, philosophical anthropology, *Lebensphilosophie* and phenomenology have been underrepresented in the sociology tradition, despite the very important critique which these traditions were able to provide in opposition to biologism, reductionism and other positivistic traditions of research.

The phenomenology of the body has been influenced by a great diversity of traditions, including *Lebensphilosophie*, philosophical anthropology and existentialism. For example, I have elsewhere (Turner 1984) attempted to show the importance of Merleau-

Ponty's work (1962) for the development of the sociology of the body. In his *Phenomenology of Perception*, Merleau-Ponty summarized much of the existing research within phenomenology and developed a brilliant conception of embodiment which attempted to overcome dualistic conceptions of mind and body. In trying to understand human perception, Merleau-Ponty argued that perception is always undertaken from a particular place or perspective. It is not possible to talk about human perception of the world without a theory of embodiment as the 'perspective' from which observation occurs. Our perception of everyday reality depends on a lived body, because, for example, we move around a room in terms of sight, touch and smell, but even our 'higher' perceptions can never escape the legacy of our (primordial) embodiment. The body is an active body which points outwards or is directed towards a habitus. Merleau-Ponty, drawing on Edmund Husserl's philosophy of intentionality, argued that a fundamental intentionality was grounded in the lived body, that is within incarnate subjectivity. Thus, perception and movement can only be artificially separated, because basic forms of perception (such as seeing itself) involve body movements. Merleau-Ponty employed psychological research on missing limbs to show how judgement and perception were radically disrupted and dislocated by injury to the body. It was as a consequence of these philosophical and psychological enquiries that Merleau-Ponty used the notion of 'embodiment' to claim that neither the Cartesian body and mind dichotomy nor the idea of body-and-mind would do justice to his contention that all 'higher' mental functions are also somatic activities. The body is not an object for itself; it is in fact 'a spontaneous synthesis of powers, a bodily spatiality, a bodily unity, a bodily intentionality, which distinguish it radically from the scientific object posed by traditional schools of thought' (Langer 1989:56).

Although both phenomenology and philosophical anthropology are somewhat remote from conventional sociological theory, it is obviously the case that anthropology as such has played a major role in attempting to conceptualize the nature of human embodiment (Turner 1991b). However, while anthropology has played an important part in maintaining a scientific interest in the social and human body, in general terms anthropology has not been concerned to understand the phenomenology of the 'lived body': it has rather been concerned to understand the body as part of a social classificatory scheme.

The idea of the body as a method of classification probably has its origins in the work of Durkheim and Mauss (1963). For example, Robert Hertz, who was one of Durkheim's students, showed how the physiological preponderance of the right side of the body had been culturally elaborated in a moral classification of good and evil. Right-handedness became a central symbol of human values (Hertz 1960). However, in contemporary thought it is in the work of Mary Douglas (1970; 1973) that we can discover an articulate understanding of the principle of the body as a symbolic system. For Douglas, the body is a metaphor of society as a whole with the consequence that disease in the body is, for example, merely a symbolic reflection of disorders in society. The stability of the body is thus a metaphor for social organization and social relationships. Our conceptual anxiety about risk and uncertainty in social relations is thereby articulated through theories of bodily order. Purity and order, sacredness and profanity, do not reside in the essence of phenomena or practices, but in their relationship to our conception of some social totality. The profane is thus disorder within the system of classificatory relationships. Douglas's ideas have been influential in anthropology, but they have also been adopted and developed in most interesting ways by sociology (O'Neill 1985; 1989).

RESCUING THE BODY

In general terms, therefore, sociology has neglected the problem and importance of the body in social life, but I have identified a number of traditions in philosophical anthropology, *Lebensphilosophie*, anthropology and phenomenology which have taken the question of human embodiment seriously. Within mainstream sociology, probably the only social theoretical tradition which seriously considered the nature of human embodiment in micro interaction was the tradition of symbolic interactionism, of which Erving Goffman was a leading exponent (1964; 1967). I have already argued that Goffman's notion of the presentation of self (1959) depended upon the view that the social self was at least in part presented through the social body. For example, the sense of embarrassment is often connected with changes in the colour of the face. In broad terms, our social notion of ease or comfort is often expressed through various bodily gestures which can be read as a sort of language.

Against this traditional absence of the body in social theory there has been, in the last decade, a significant number of influential studies which have begun to take the sociology of the body seriously (Frank 1991). In the area of medical history, Thomas Laqueur (1990), Emily Martin (1987), Aline Rousselle (1988), Philippe Ariès and André Béjin (1985), and David Armstrong (1983) have shown how significant an understanding of the body is to a sophisticated history of medical knowledge and medical power. This growing appreciation of the importance of the body has also now begun to influence the way in which the history of Christian theology is to be written (P. Brown 1988). The influence of Heidegger's concern for ontology is now also influential in the philosophical analysis of the idea of the lived body (Levin 1985; 1988). One could identify similar developments in philosophy (Hudson 1982; Shapiro 1985). Much recent work on Heidegger has in fact contributed to a richer philosophical awareness of the human body. In this respect, Jacques Derrida's exegesis of Heidegger's work is especially fruitful (Derrida 1987). Clearly, contemporary feminist thought has played a major role in re-establishing questions of body, gender and sexuality on the agenda of social theory (Suleiman 1986). These feminist critiques have problematized the conventional distinctions in classical social theory between nature and culture, specifically between the idea of the contrast between woman in nature and man in culture (Sydie 1987). The sociology of the body may also prove to be a significant dimension of the growing literature of the social division of emotional work in relation to patriarchy and the sexual division of labour (Hochschild 1983). Finally, a major collection of articles was edited by Michel Feher with Romona Naddaff and Nadia Tazi under the title *Fragments for a History of the Human Body* (Feher 1989). There is, therefore, a virtual industry of publishing on the topic of the body which has gained momentum within the last few years. How might we explain the return of the body to the social theoretical gaze?

In providing an account for recent interest in the body I shall briefly consider four aspects of contemporary society which bear upon the problem of the body in relation to personality, nature and culture. Feminist theory has been fundamental to contemporary theories of the body because the feminist movement has problematized the relationship between biology, gender and sexuality (Wallace 1989). Rather than attempting to cover the complex and extensive literature on feminism and the body (Frank 1990, 1991;

Fraser and Nicholson 1989; Rosaldo and Lamphere 1973; Suleiman 1986; Wallace 1989), I shall briefly comment on Donna Haraway's discussion of cyborgs, that is cybernetic organisms, which, in her view, raise problems for the whole set of distinctions which have existed between nature and society (Haraway 1989). For Haraway, the myth of the cyborg has complicated the relationship between nature and the machine, because the cyborg crosses and confuses the boundaries which have become conventional in technological history. She argued that 'the cyborg is a creature in a postgender world; it has no truck with bisexuality, pre-oedipal symbiosis, or other seductions to organic wholeness through a final appropriation of the parts into a higher unity' (Haraway 1989: 192). The cyborg opens up a series of 'leaky distinctions' between nature, society and culture. The cyborg crosses the boundary between the animal/human organism and the technical machine, but it also brings into question the boundary between the physical and the non-physical world. We can elaborate Haraway's argument by suggesting that major changes in medical technology, in particular in relation to human reproduction, have brought into question the nature of sexual identity. There has been, therefore, a simultaneous questioning of the nature of the body in relation to gender in politics and culture alongside a technical revolution in medicine which has brought into question the nature of human reproduction as such.

This political and technological struggle around the body was fundamental to Michel Foucault's argument in volumes one, two and three of *The History of Sexuality* (1979, 1987, 1988) that contemporary politics was bio-politics. The state is increasingly important in the regulation of human bodies through medical legislation concerning such things as abortion, child-care, IVF programmes, the regulation of AIDS as a modern epidemic, legislation concerning citizenship rights in terms of sex changes, the state's regulation of surrogate parenthood and so forth. The politicization of the body has brought into focus the complex interrelationship between citizenship, embodiment and gender. These broad changes in the politics of sex is one set of empirical processes which lies behind the contemporary emergence of the body in social theory as a topic.

The politicization of the body and the feminization of life have contributed to an interest in the social analysis of human embodiment. These two related social changes should also be connected with developments in contemporary consumerism. The growth of a consumer culture and the fashion industry in the twentieth century have given a special emphasis to the surface of the body (Featherstone *et al.* 1991). In the growth of a consumer society with its emphasis on the athletic/beautiful body, we can see a major historical transformation of Western values from an emphasis on the internal control of the body for ascetic reasons to the manipulation of the exterior body for aesthetic purposes (Turner 1984). This transformation of the body represents a secularization of Western values in which diet for the management of the spirit and the life of the soul has been transformed into a diet for the purposes of longevity and sexuality. The original formula of dietary management for a government of the body has been therefore converted through medicalization into a secular morals of fitness and hygiene.

The growing emphasis on the aesthetic quality of the body in relation to consumerism has emphasized the virtues of thinness and self-regulation in the interests of looking good. The body has become a fundamental feature of taste and distinction in which the management of the human form becomes part of the major aspect of cultural or physical capital (Bourdieu 1984). Although these changes are very general in society, there are good reasons for believing that they have a special impact on the new middle classes, that is on the urban culture which may be associated with the end of organized capitalism and with postmodern city culture. Although this argument is clearly contentious, it is certainly the case that different social classes develop different body images and, according to Bourdieu, while the middle class prefers fitness, the working class develops bodies which exhibit male strength. We can thus see that the body is brought into fashion and consumer culture as a mark of distinction, as a symbolic representation of class differences, as a field for gender differentiation, and as a potentiality which must be managed in the process of ageing in order for the individual to remain part of the scene.

The fragmentation of consumer culture, the differentiation of style and taste by social strata, the development of leisure centres and city culture in the era of disorganized capitalism are changes which have promoted a debate, firstly about the idea of postmodern culture as such (Turner 1990), but secondly about the possibility of a postmodern body of which the cyborg would be one example. There is a growing awareness that the body is socially produced and socially constructed, and that the body is fragmented and has many identities, and that the body is no longer secured or located in some fixed social space. A postmodern culture is characterized by panic and the body has become a target of many invasions (Kroker and Kroker 1987).

As the body has become fashionable and codified, so there has been a growing emphasis in social theory on desire, sexuality and emotion which is part of the poststructuralist movement which has dominated much of the thought of Michel Foucault, Jacques Derrida and Jean Baudrillard. This poststructuralist turn in social theory can itself be seen as part of the contemporary critique of the Cartesian legacy of the modernist movement, which had its origins in seventeenth-century science and colonial capitalism. Although the body has become increasingly prominent in contemporary debate, the nature of the body has become theoretically a complex issue which we must attempt to unravel.

ANALYTICAL FRAMEWORKS

In this discussion of analytical approaches to the human body, I am specifically not concerned with natural science approaches; that is, I am not concerned with the body as an organism within the framework of the natural and medical sciences. My concern is to discuss the approaches which have developed within social sciences or which have a direct implication for social science. Broadly speaking, within the social sciences, we are faced with the choice in terms of ontology between foundationalist and anti-foundationalist perspectives on the body. Foundationalist frameworks are concerned to understand the body as a lived experience, or to comprehend the phenomenology of embodiment, or to understand how the biological conditions of existence impinge upon the everyday life and macro organization of human populations, or they want to understand how the historical demography of societies has influenced the course of human history, or they seek to analyse the complex interaction between the organic systems, cultural frameworks and social processes. By contrast, anti-foundationalist perspectives conceptualize the body as a discourse about the nature of social relations, or comprehend the body as a system of symbols, or seek to understand how bodily practices are metaphors for larger social structures, or they understand the body as a social construction of power and knowledge in society, or perceive the body as an effect of social discourse. Within these two perspectives, there are, as I have indicated, many different approaches and schools, but as a general organizing principle, sociological approaches to the body tend to divide along this philosophical issue as to the ontological status of the body. We also have to recognize that any specific author might inconsistently or covertly employ several philosophies of ontology simultaneously.

These questions about ontology are parallel to questions about epistemology in sociology. In epistemological terms, the principal debate has been between social constructionists and anti-constructionists. For those who oppose the idea of the social construction of reality, the body exists independently of the forms of discourse which represent it; for their opponents, the body is socially constructed by discursive practices. As a result, there is a significant debate about whether these epistemological orientations to the body are actually compatible or mutually exclusive.

This epistemological division is also often associated with a modern and postmodern orientation, since anti-foundationalist postmodernism wants to deconstruct existing discourses about the body in order to demystify the concept of the body from its existing traditional paradigms. In this preliminary discussion, I shall argue that we do not in fact have to choose between these competing orientations, because some degree of theoretical reintegration and rapprochement may be possible. My concern is to try to establish some minimal theoretical synthesis that will permit and promote a diverse tradition of social theory within a common framework of interest in the body. My own interpretation is that anti-foundationalist approaches are in fact concerned with rather different issues and topics, and that they are addressed to rather different analytical questions. Therefore, they do not present mutually exclusive positions on the human body.

In this account of approaches, I shall start with those approaches in which the body is primarily conceptualized as a symbolic system. The notion that the body is a system of communication is a well established position within the humanities and social sciences. For example, many of the metaphors by which we speak about politics and society are based upon the body – such as the body politic, the head of state, the arms of the church or the body of the church. Our view of these symbolic properties of the body depends a great deal upon the brilliant research of Ernst Kantorowicz in his *The King's Two Bodies* (1957) where Kantorowicz provided an illuminating analysis of the historical development of political sovereignty as it was expressed in theories of the king's embodiment. Whereas kingship originally resided in the physical body of the king, with the evolution of political theory and institution of power there emerged a separation between the king's physical body and symbolic body. The symbolic body of the king came eventually to represent abstract sovereign power, and hence it was thought that the king had a physical corruptible body and an abstract divine body. It was the symbolic body of the king which guaranteed the continuity of sovereign state power despite the periodic death of particular kings. Thus, upon the death of the king it was common for the courtiers to shout 'The king is dead, long live the king!' Because the symbolic wholeness of the king's body was particularly important to the continuity of state power, attacks upon the king were regarded as attacks upon the nation state. Michel Foucault in his *Discipline and Punish* (1977) makes a great play upon the Kantorowicz theory of the king's body in the opening account of the treatment of regicides in traditional French political culture. An attack on the king's body was an attack on society itself and therefore the punishment of regicides had to correspond fairly directly to the seriousness of a political crime, so that violent punishment of the body was a necessary form of state violence. According to Foucault, we can write the history of Western penal regulations in terms of a transition from the violent retribution of society on the body of the criminal to the disciplined management and regulation of docile bodies within the system, which was originally developed by Jeremy Bentham under the notion of panopticism. The punishment of the body under the scaffold in the interests of social revenge was eventually replaced by the discipline of the penitentiary as the primary moral mechanism for the management of deviance.

In the history of the body in medical sciences, it is interesting to note that the surgical manipulation of the body in the anatomy lesson created enormous moral and religious problems, because in opening up the body the surgeon was opening up the mystery of the universe. What God had closed within the body should not be opened up for secular purposes by the surgeon. The surgeon's exposure to the internal fluids and juices of the body, particularly the blood, also exposed him to moral and spiritual dangers. Medieval regulations for the control of surgery often recommended moral practices for the surgeon prior to an operation which were parallel to the preparations a priest undertook before giving the sacrament (Pouchelle 1990).

From the point of view of a sociology of the body, therefore, the history of the anatomy lesson is particularly instructive in understanding the place and province of the body in human societies. I have claimed elsewhere (Turner 1987) that within the Western cultural tradition of Christianity there developed three major institutional patterns or responses to the earthliness or fleshliness of our existence, namely the spheres of religion, law and medicine. Prior to modern social differentiation, these spheres were not separate and they can be regarded consequently as institutional superstructures which have been organized around and in response to the spiritual dilemmas of human embodiment and the need for cultural management of this embodiment. In taking this position I am partly following Gehlen's theory (1988) that we can understand religion, for example, as a directional system that permits conscious and active adjustment to the world. Medicine, religion and law are social arrangements which are societal responses to the embodiment of human beings in the world and the reciprocal connections in everyday life between such embodied persons. The anatomy lesson in the seventeeth century therefore was a medical, legal and religious practice in which the body of a criminal was opened up to public inspection and to the moral gaze of society. Thus the public anatomical dissection of cadavers was part of a broader juridical management of social deviance.

I have already shown how the idea that the body is a symbolic system has come to its main focus in the anthropological work of writers like Mary Douglas. This tradition is well known in the social sciences, and Mary Douglas's work is sufficiently well known in general not to require detailed commentary here. It will suffice to say that Douglas's work is in fact about the nature of risk in human societies and social responses to risk, where the body provides a metaphor of coherence and disorder. In this specific sense, Douglas's anthropology is not an anthropology of the body, but an anthropology of the symbolism of risk. The idea that the body is a metaphor of social arrangements has of course continued to influence feminist theory, of which Emily Martin's *The Woman in the Body* (1987) is a particularly good example.

Martin's study is an important contribution to the anthropology of the body, but her main focus is still the question of the representation of the body in medical and other discourse. She points out correctly that, while we are prone to see the medical language of previous centuries as a symbolic representation of social ideas, we are unwilling to treat contemporary scientific representations as anything other than natural descriptions of the physical world.

While Laqueur in *Making Sex* (1990) has shown that the anatomical representation of men and women from Galen onwards directly reflected social attitudes about the inequality of women as inferior reproductive systems, we are less likely to think about modern medical representations in the same anthropological framework. Martin's approach in this context is important in the ways in which she has discovered how in our technological society we tend to see the body in terms of contemporary electrical metaphors. The imagery of the contemporary biochemistry of the human cell has often been that of the factory in which the cell functions as a specialized mechanism for the conversion of energy. Alternatively, the human organism is seen in terms of economic metaphors of energy and production. Argumentation in molecular biology has often been conducted in terms of metaphors of an information science, concerning management and control. For example, the flow of information between DNA and RNA results in the production of protein. In speaking about birth as labour we often forget the implicit economic metaphors of such birth imagery.

We can see in this literature a rich tradition of anthropological analysis of the symbolism of the body where the body is treated as a symbolic system or as a discourse. These traditions by and large are not interested in the physiological body and they are equally uninterested in the notion of the lived body. Perhaps one of the most significant contributions to the sociology of the body in recent years has, however, come from the work of Foucault who is equally unimpressed by the phenomenological tradition. In part, Foucault's tendency to see the body as an effect of discursive power results from his rejection of Heidegger, whose theory of being is inimicable to Foucault's approach. Heidegger's fundamental ontology in *Being and Time* (1962) attempted to ground any philosophical discussion of being in the facticity of everyday existence. Foucault turned his back on such an approach to being, by arguing in *The Order of Things* (1970) that all understanding is constrained within and produced by the frameworks of epistemology which happened to be dominant within any given period. The representation of reality is thus an effect of an episteme which controls and regulates the way in which conceptualization can occur. Foucault's approach appears to reject the facticity of the body, an idea which was fundamental to Heidegger, by saying that the body is produced by knowledge or that the body is an effect of practices which embody such forms of knowledge. His research has been concerned with how 'bodies' are produced by discourses and his primary theme was the normalization of the body and populations by the social sciences and the institutions which articulated scientific knowledge. This work was fundamental, but it was not concerned with the nature of the body as *Leib*. Foucault's work, therefore, appears to reject the idea of a universal ontology and thus rejects any attempt to think about the body as a grounding for such a universalism. The body in Foucault's early work is, so to speak, made possible by the emergence of natural sciences, such as biology, physiology and chemistry. Concepts such as the body or populations are components within a discursive framework which makes it possible to think about bodies and populations. However, Foucault's main purpose was not to create an epistemology of the social sciences, but to understand the complicated relationship between power, truth and knowledge in Western thought.

In his early work, it appears as if Foucault was studying how the body can emerge in different practices, which are related to the control and management of human beings. Thus, in *The Birth of the Clinic* (1973), Foucault was concerned to see how medical knowledge and practice produced the body and appropriated it within a network of institutions which functioned at the micro level to establish medical power. Similarly, in his study of the prison, Foucault (1977) analysed the growth of the disciplined and docile body as an effect of penitentiary practices which were linked to a utilitarian theory of pain. In his study of the history of sexuality, Foucault (1979) considered how the emergence of a discourse of sex in the nineteenth century produced sex as a topic and thus how sex became an object of political struggle exercised through particular medical knowledge.

While Foucault's work appears to be anti-foundational in its epistemology, there is a theme of romanticism in Foucault's outlook in which the primitive body, existing before signification, represents a world of innocent enjoyment. There is some element of truth in Richard Rorty's condemnation of Foucault as an immature anarchist but Rorty does find something valuable in Foucault once we have set 'aside all the anarchist claptrap about repression and all the Nietzschean bravura about the will to power' (Rorty 1986:47). The truth with which Foucault was struggling was that ultimately all conceptual perspectives on the world are incommensurable. In Foucault we find a refusal to make easy moral and political judgements on reality, because it is always difficult to avoid existing power positions. One description of the world may be as good or as bad as any other. Secondly, Foucault was engaged in rejecting the bureaucratic rationalization of the world which produces 'the overwhelming sameness of it all' (Boyne 1990:139). Foucault rejected the detailed normalization of the body which has been an effect of modern rationalization. However, if Foucault wanted a political and moral programme, he had to find a foundation upon which to launch such a critique. My own view is that, behind Foucault's critique, there lurked a quest for the Other in the untrammelled, uncivilized, prediscursive body. The nostalgia in Foucault's philosophy was the search for the Sexual Body before the social contract.

Perhaps the final anthropological contribution to the debate about the body can be illustrated by the work of Pierre Bourdieu. Although it may seem strange to define Bourdieu as an anthropologist, his early fieldwork and his theory of practice are quite clearly developed within an anthropological perspective (Bourdieu 1977). Bourdieu did his early anthropological research on the Kabyle; he developed an anti-structuralist anthropological position which attempted to correct a number of problems which he identified in the work of Claude Lévi-Strauss. It was this critique of structuralism which provided the background to his book *Outline of a Theory of Practice* (1977).

Bourdieu became famous, of course, for his later contribution to educational sociology where, in developing the work of Marx, he distinguished cultural capital from social capital (Bourdieu and Passeron 1977). Society is seen as an organization of different fields which are the sites of individual and group struggle over the production and consumption of capital goods. The value of a symbolic commodity is determined by the quantity of symbolic capital which a producer has accumulated. Success in social confrontations permits a dominant social class to exercise symbolic violence over other consumers within its cultural field.

Bourdieu's sociology is provocative because it attempts to show how in the world of high culture similar struggles for dominance are undertaken. Bourdieu is of interest to us because he has developed (a largely implicit) sociology of the body as part of his more general concern with the notion of habitus and practice. In *Distinction* (Bourdieu 1984) the symbolic representation of the body and the dispositions of the body in respect of taste are an important feature of his concept of cultural capital. The human body in Bourdieu's sociology appears as a site or space on which is inscribed the cultural practices of various social classes. Each class and each class fraction has a characteristic activity, especially sport, which exhibits both their economic and cultural aspects. The body in Bourdieu's theory can be regarded as a carrier of class dispositions which are themselves the channels of interests within the habitus or life-world of various classes. It is important to make clear that Bourdieu's work obviously assumes the existence of the organic human body, but he argues that this 'raw material' is shaped and constructed by social class forces; the body becomes part of the cultural capital of an individual, and the body in this sense is a sign of power.

The principal alternative to anti-foundationalist approaches to the body, which place a special emphasis on discourse, may be found in the phenomenological tradition, in philosophical anthropology and in anthropology generally. This tradition is primarily concerned with the concept of the lived body. I have already indicated the importance of the work of Gehlen, Berger, Heidegger, Merleau-Ponty and Plessner. Rather than attempting to survey this field as a whole, I shall select one particular rather neglected contribution by the phenomenologist Schilder for commentary. Paul Schilder (1886–1940) published *The Image and Appearance of the Human Body* in 1935 (Schilder 1964). Schilder's book was divided into three sections, namely the physiological basis of the body image, the libidinous structure of the body image and finally the sociology of the body image. He was concerned to describe and to analyse what he called the 'postural model' of the body which he thought was a constructed image which was only indirectly related to either the physiological or purely sensory character of the human body. He illustrated his idea by reference to a wide range of pathological findings such as aphasia and brain lesions in order to illustrate the idea of the conventional nature and constructive character of the normal body consciousness. By the body's libidinous structure, Schilder had in mind the emotional, feeling tone of the body. Whereas the postural model of the body was primarily concerned with the external bodily organization of space, the libidinal structure of the body was concerned with the interior ordering of body time. In the final section of the book on sociology of the body, he attempted to illustrate the social character of the body image; Schilder believed that the body image is necessarily social and that all aspects of the body image are constructed and developed through social relations. Thus Schilder wrote that 'body images are in principle social. Our own body image is never isolated but is always accompanied by the image of others' (Schilder 1964:240–1). Schilder's work went a long way to integrating psychological, sociological and cultural understanding of the body image as a fundamental aspect of personality and social interaction. He wrote that 'there is no body-image without personality. But the full development of the personality of another and its values is only possible through the medium of the body and the body-image. The preservation, construction and building-up of the body-image of this other thus become a sign, signal, and symbol for the value of his integrated personality' (Schilder 1964:282). Finally, Schilder followed Max Scheler in arguing that we should not treat the objective body (*Körper*) as a separate entity from the inner sensations of the subjective body (*Leib*). He argued that 'there is only one unit. It is the body, and there is an outside of the body and a substance of heavy mass which fills this body. But it is true that the body in this sense is always present; it is not the product of sensations, which get their final meaning only from the unit which is one of the fundamental units of our experience' (Schilder 1964:283).

We find similar ideas about the phenomenology of the body in the work of philosopher Merleau-Ponty (1962) who quoted from some of Schilder's earlier research, namely *Das Körperschema* (1923). Merleau-Ponty also provides us with a way of thinking about the lived body in relation to the organic natural world. Again, by using a conventional distinction in German, Merleau-Ponty wrote that man 'has not only a setting (*Umwelt*), but also a world (*Welt*) (Merleau-Ponty 1962:87). In trying to give a more precise meaning to the notion of embodiment, he went on to argue that

> Man taken as a concrete being is not a psyche joined to an organism, but movement to and fro of existence which at one time allows itself to take corporeal form and at others moves towards personal acts. ... It is never a question of the incomprehensive meeting of two casualties, nor of a collision between the order of causes and that of ends. But by an imperceptible twist an organic process issues into human behaviour, an instinctive act changes direction and becomes a sentiment, or conversely a human act becomes torpid and is continued absent-mindedly in the form of a reflex.
>
> (Merleau-Ponty 1962:88)

Sociological theory is often written as if one had to choose between competing and incommensurable paradigms. My own view, which could be called methodological pragmatism, is that the epistemological standpoint, theoretical orientation and methodological technique which a social scientist adopts, should be at least in part determined by the nature of the problem and by the level of explanation which is required. For example, the study of queueing in traffic jams and the study of anorexia nervosa are both topics which are appropriate objects for research by sociology, but they do not necessarily raise the same order of epistemological, ontological or theoretical questions. While queueing in traffic jams may raise few fundamental philosophical problems for a sociologist, the nature of a disease entity is highly complicated philosophically and requires a great deal of analytical clarification (King 1982). As a consequence, I see no compelling theoretical reason for opting categorically for a position in which the body is treated as either socially constructed and discursive or as a lived body from within a phenomenological or philosophical anthropological perspective. There appear to be strong reasons for regarding the body as simultaneously both discursive and animated, both *Körper* and *Leib*, both socially constructed and objective. The emphasis which we give to any or all of these dichotomies will depend on what type of research we want to undertake.

Thus, in social life, it seems to be clear that bodies and their diseases are deeply metaphorical. This metaphoricality has been particularly underlined by social responses to the nature of AIDS (Sontag 1988). At the same time, it seems equally obvious that there is an important theoretical space for a phenomenology of disease in which the social researcher is concerned with the lived experience of pain, discomfort and alienation. In the work of Oliver Sacks, we have been presented with eloquent phenomenological insights into the human experience of the alienating consequences of disease and illness in his sympathetic studies of Parkinson's disease (Sacks 1976), migraine (Sacks 1981) and physical injury (Sacks 1984). I can see no good purpose in adopting epistemological fundamentalism in order to achieve some analytical purity wherein we could conceptualize the body within a single philosphical paradigm. By contrast we should encourage research which will be open both to the idea of the body as lived experience (*Leib*) and to the discourse of the body as an objective presence (*Körper*). The disorders of the body require a language by which we might both describe and experience them; for example, in the English language at least, diseases are often described from within a military metaphor. Because we are attacked by viruses, we have to be on guard against infection. There is a campaign against AIDS. At the same time we want to describe pain as indescribable. The phenomenological misery of pain is literally beyond language and beyond speech. In this sense it is a sort of transgression.

THE SOCIOLOGY OF THE BODY

Partly influenced by the work of Foucault, I have elsewhere suggested (Turner 1984:91) that we might usefully think about the body in sociological terms as divided into an internal and external space. The external character of the body is concerned with the representations of the bodies in social spaces and their regulation and control. To a large extent, the sociology of the body (in its as yet underdeveloped form) has been primarily concerned with the question of how the body is represented in space in relation to personality and identity. The work of Goffman in this respect is exemplary. The externalities of the body have become a popular topic of research in the sociology of consumer culture, for example (Featherstone 1982).

This interest in the external body thereby contrasts neatly with research which we might suggest is concerned with the internal structure, organization and maintenance of the body. Whereas the external problem is the problem of representation, the interior problem of the body is one of restraint, that is of the control of desire, passion and need in the interests of social organization and stability. The work of Nietzsche, Weber and Freud on the management of desire can be seen as contributions, from various philosophical, sociological and psychological standpoints, to an analysis of interior maintenance of the body. Within this typology I went on to suggest that rather than looking only at bodies in the singular, we might think à la Foucault of bodies in the plural, that is of populations. I suggested therefore that the two corresponding problems of the population were those of reproduction and regulation. That is, populations have to be reproduced through time through the control of sexuality within the household and within the family. At the same time, particularly with the growth of urban congestion and the social problems of city life, much social theory had revolved around the problem of social regulation. These

	Populations	Bodies	
Time	Reproduction Patriarchy	Restraint Asceticism	Internal
Space	Regulation Panopticism	Representation Commodification	External

Figure 1   Societal task model

dimensions (internal/external and bodies/populations) produce a four-fold table of reproduction, restraint, representation and regulation (see Figure 1). This typology has been usefully modified and extended by Arthur Frank (1991) who has argued that these institutional sub-systems may also be defined or specified by reference to patriarchy, asceticism, panopticism and commodification.

Having commented on this typology, Frank in his elaboration of my argument went on to suggest some important and new additions to this approach to the body which I wish to discuss here. Frank's objection to my position is that I, as it were, approached the problem of the body from the perspective of society, and thus I moved theoretically downwards towards the body from the level of the societal, whereas an alternative and perhaps prior orientation would be to start with the body's problems for itself. He argues that 'I propose instead to begin with how the body is a problem *for itself*, which is an action problem rather than a system problem, proceeding from a phenomenological orientation rather than a functional one. If theoretical objective is to move from the body to self to society, Turner's typology would represent final, societal level of theorising' (Frank 1991:47-8). He goes on to argue that only bodies can truly be said to have 'tasks'. Adopting a version of Giddens's structuration theory, Frank argues that we should see the body as both the outcome and the medium of social practices, namely of body techniques. He thus suggests that bodies exist between discourses and institutions, where discourses signify the mappings of the body's possibilities and limitations. These mappings provide the normative paradigm within which the body can as it were understand itself. By contrast, institutions are the places or contexts within which these practices occur. He claims that 'we must recognise institutions from the beginning since the actions of bodies are already oriented to institutional contexts' (Frank 1991:49). Of course, Frank also wants to assert that physiology or more specifically corporeality exists as a third dimension of the constitution of the body. Thus, 'beyond the relative discourse of physiology, corporeal reality remains an obdurate fact. There is a flesh which is formed in the womb, transfigured (for better or worse) in its life, dies and decomposes. Thus what I am calling ''the body'' is constituted in the intersection of an equilateral triangle the points of which are institutions, discourses and corporeal reality' (Frank 1991:49).

Thus we might take as an example the ascetic practices of the medieval period which were designed to regulate and produce the spirituality of bodies. In this example, the institution is clearly the medieval church with its complex set of roles and practices whereby ecclesiastical institutions routinized charisma. The discourses which are relevant to these ascetic practices were the discourses of diet (Turner 1982). These discourses specify the goals, objectives and boundaries within which the aesthetic regulation of desire was to take place. These discourses are well illustrated by, for example, Rousselle's *Porneia* (1988). However, in order to write this history we need to look beyond both institutions and discourses to the question of the corporeality of the body itself. We need to pose the question of how much self-punishment, deprivation and asceticism this medieval corporeality could sustain. We have to realize that the body itself might have a history (Bynum 1987).

CONCLUSION

In this discussion, I have outlined both why the sociological study of the body is important and why the study of the body is complex. In terms of the importance of the body, recent social, cultural and technological changes have made the body central to modern politics, because the conventional boundaries between the natural and the social are constantly eroded and changed. Political positions are as a result rapidly rendered obsolete. What are the limitations on the individual in the modern world, especially in bioethical questions? Thus, 'the very people who ten years ago were militating for the right to abortion are appalled by the prospects for genetic manipulation, the plan of painlessly ending the life of abnormal newborns, or the inextricable legal and emotional conflicts stirred up by surrogate motherhood' (Ferry and Renaut 1990:18). The fictional nightmare of *Robocop* is a reality which, as we all know, is only around the corner.

How we study the body is equally complex, because the number of competing traditions appears to be infinite. In order to simplify the argument, I have suggested that ontologies of the body tend to bifurcate around foundationalism and anti-foundationalism: is the fundamental nature of the body produced by social processes, in which case the body is not a unitary or universal phenomenon, or is the body an organic reality which exists independently of its social representation? Similarly, we can say that epistemologies of the body are divided between social constructionists and anti-constructionists. The body is a product of knowledge which cannot exist independently of the practices which constantly produce it in time and space. Alternatively, the body exists separately from its social construction. Within the framework of these dichotomies, I have examined philosophical anthropology, phenomenology, anthropology, history and sociology as disciplines which have contributed to the analysis of the body.

My own strategy has been one of epistemological pragmatism. If we are interested in the social representation, for example of the reproductive organs, then it makes sense to think of 'the body' as a representation of power. If we are interested in how a missing limb has an impact on body image, then the psychological research of Paul Schilder may be more relevant than in the approach of Foucault. Given the fact that a sociology of the body as a topic in mainstream sociology is relatively new, it is inappropriate to foreclose our conceptual options prematurely.

NOTE

1   In discussing the work of the philosophical anthropologists, I have retained their terminology. Although their work is by modern standards sexist, I cannot see that much is to be gained by updating their terminology to give them a false liberalism, let alone a false feminism. When Arnold Gehlen talked about 'Man', he was not persuaded that an alternative formula might have been appropriate. The recent translation of his *Der Mensch* by Columbia University Press was published as *Man* (Gehlen 1988). The translators' Preface makes the point similarly that 'to avoid sexist terminology would add a contemporary cast to the text that, in our opinion, would not be appropriate' (Gehlen 1988: vii). Where I have been expressing my own opinions, I have tried to avoid a sexist vocabulary by using the expression 'humanity' or 'human beings'.

REFERENCES

Abercrombie, N., S. Hill and B.S. Turner (1986) *Sovereign Individuals of Capitalism*, London: Allen & Unwin.

Archer, M. (1988) *Culture and Agency, the Place of Culture in Social Theory*, Cambridge: Cambridge University Press.

Ariès, P. and A. Béjin (eds) (1985) *Western Sexuality, Practice and Precept in Past and Present Times*, Oxford: Basil Blackwell.

Armstrong, D. (1983) *Political Anatomy of the Body, Medical Knowledge in Britain in the Twentieth Century*, Cambridge: Cambridge University Press.

Berger, P.L. (1969) *The Social Reality of Religion*, London: Faber & Faber.

Berger, P.L. and T. Luckmann (1967) *The Social Construction of Reality*, London: Allen Lane.

Bourdieu, P. (1977) *Outline of a Theory of Practice*, Cambridge: Cambridge University Press.

Bourdieu, P. (1984) *Distinction, a Social Critique of the Judgement of Taste*, London: Routledge & Kegan Paul.

Bourdieu, M. and J-C. Passeron (1977) *Reproduction in Education, Society and Culture*, London: Routledge & Kegan Paul.

Boyne, R. (1990) *Foucault and Derrida, the Other Side of Reason*, London: Unwin Hyman.

Brown, P. (1988) *The Body and Society. Men, Women and Sexual Renunciation in Early Christianity*, New York: Columbia University Press.

Brown, T. (1985) 'Descartes, dualism and psychosomatic medicine', pp. 40–62 in W.F. Bynum, Roy Porter and Michael Shepherd (eds) *The Anatomy of Madness, Essays in the History of Psychiatry*, London: Tavistock.

Bynum, C.W. (1987) *Holy Feast and Holy Fast, the Religious Significance of Food to Medieval Women*, Berkeley: California University Press.

Deleuze, G. (1983) *Nietzsche and Philosophy*, London: The Athlone Press.

Derrida, J. (1987) '*Geschlecht* II: Heidegger's Hand', pp. 161–96 in J. Sallis (ed.) *Deconstruction and Philosophy, the Texts of Derrida*, Chicago and London: Chicago University Press.

Douglas, M. (1970) *Purity and Danger, an Analysis of Concepts of Pollution and Taboo*, Harmondsworth: Penguin Books.

Douglas, M. (1973) *Natural Symbols, Explorations in Cosmology*, Harmondsworth: Penguin Books.

Durkheim, E. and M. Mauss (1963) *Primitive Classification*, Chicago: Chicago University Press.

Featherstone, M. (1982) 'The body in consumer culture', *Theory, Culture & Society* vol. 1(2):18–33.

Featherstone, M., M. Hepworth and B.S. Turner (eds) (1991) *The Body, Social Process and Cultural Theory*, London: Sage.

Feher, M. with R. Naddaff and N. Tazi (1989) *Fragments for a History of the Human Body*, New York: Zone, 3 vols.

Ferry, L. and A. Renaut (1990) *Heidegger and Modernity*, Chicago and London: Chicago University Press.

Foucault, M. (1970) *The Order of Things, an Archaeology of the Human Sciences*, London: Tavistock.

Foucault, M. (1973) *The Birth of the Clinic*, London: Tavistock.

Foucault, M. (1977) *Discipline and Punish, the Birth of the Prison*, London: Tavistock.

Foucault, M. (1979) *The History of Sexuality; Volume One: An Introduction*, London: Tavistock.

Foucault, M. (1987) *The History of Sexuality; Volume Two: The Use of Pleasure*, London: Allen Lane.

Foucault, M. (1988) *The History of Sexuality; Volume Three: The Care of the Self*, London: Allen Lane.

Frank, A. (1990) 'Bringing bodies back in: a decade review', *Theory, Culture & Society* vol. 7(1):131–62.

Frank, A. (1991) 'For a sociology of the body, an analytical review', pp. 36–102 in M. Featherstone *et al.* (eds) *The Body, Social Process and Cultural Theory*, London: Sage.

Fraser, N. and L. Nicholson (1989) 'Social criticism without philosophy, an encounter between feminism and postmodernism', *Theory, Culture & Society* vol. 5(2–3):373–94.

Gehlen, A. (1988) *Man, His Nature and Place in the World*, New York: Columbia University Press.

Giddens, A. (1984) *The Constitution of Society*, Cambridge: Polity Press.

Goffman, E. (1959) *The Presentation of Self in Everyday Life*, Garden City, New York: Doubleday Books.

Goffman, E. (1964) *Stigma, Notes on the Management of Spoiled Identity*, Englewood Cliffs, New Jersey: Prentice-Hall.

Goffman, E. (1967) *Interaction Ritual, Essays in Face-to-Face Behavior*, Chicago: Aldine.

Haraway, D. (1989) 'Manifesto for Cyborgs, science, technology, and socialist feminism in the 1980s', pp. 190–234 in L. Nicholson (ed.) *Feminism and Postmodernism*, London: Routledge.

Heidegger, M. (1962) *Being and Time*, Oxford, Basil Blackwell.

Heidegger, M. (1982) *Gesamtausgabe, band 54, Parmenides*, Frankfurt am Main: Vittorio Klostermann.

Heidegger, M. (1989) *What is Philosophy?* London: Vision Books.

Hertz, R. (1960) *Death and the Right Hand*, New York: Cohen & West.

Herzlich, C. and J. Pierret (1987) *Illness and Self in Society*, Baltimore and London: Johns Hopkins University Press.

Hochschild, A. (1983) *The Managed Heart, Commercialization of Human Feeling*, Berkeley: University of California Press.

Holton, R.J. and B.S. Turner (1986) *Talcott Parsons on Economy and Society*, London: Routledge & Kegan Paul.

Honneth, A. and H. Joas (1988) *Social Action and Human Nature*, Cambridge: University of Cambridge.

Hudson, L. (1982) *Bodies of Knowledge, the Psychological Significance of the Nude in Art*, London: Weidenfeld & Nicolson.

Kantorowicz, E.H. (1957) *The King's Two Bodies*, Princeton, NJ: University of Princeton Press.

King, L.S. (1982) *Medical Thinking: a Historical Preface*, Princeton, NJ: Princeton University Press.

Kroker, A. and M. Kroker (eds) (1987) *Body Invaders, Panic Sex in America*, Montreal: New World Perspectives.

Langer, M.M. (1989) *Merleau-Ponty's Phenomenology of Perception, a Guide and Commentary*, London: Macmillan Press.

Laqueur, W. (1990) *Making Sex, Body and Gender from the Greeks to Freud*, Cambridge, Mass: Harvard University Press.

Leder, D. (1990) *The Absent Body*, Chicago and London: Chicago University Press.

Levin, D.M. (1985) *The Body's Recollection of Being, Phenomenological Psychology and the Deconstructionism of Nihilism*, London: Routledge & Kegan Paul.

Levin, D.M. (1988) *The Opening of Vision, Nihilism and the Postmodern Situation*, New York and London: Routledge.

Martin, E. (1987) *The Woman in the Body, a Cultural Analysis of Reproduction*, Milton Keynes: The Open University Press.

Mauss, M. (1979) *Sociology and Psychology, Essays*, London: Routledge & Kegan Paul.

Mead, G.H. (1934) *Mind, Self and Society: From the Standpoint of a Social Behaviorist*, Chicago: Chicago University Press.

Merleau-Ponty, M. (1962) *Phenomenology of Perception*, London: Routledge & Kegan Paul.

Nehamas, A. (1985) *Nietzsche, Life as Literature*, Cambridge, Mass: Harvard University Press.

O'Neill, J. (1985) *Five Bodies, the Human Shape of Modern Society*, Ithaca, NY: Cornell University Press.

O'Neill, J. (1989) *The Communicative Body, Studies in Communicative Philosophy: Politics and Sociology*, Evanston: Northwestern University Press.

Parsons, T. (1937) *The Structure of Social Action*, New York: McGraw-Hill.

van Peursen, C.A. (1956) *Lichaam-Ziel-Geest, inleiding tot een fenomenologie anthropologie*, Utrecht: Erven J. Bijleveld.

Plessner, H. (1976) *Die Frage nach der Conditio humana, aufsatze zur philosophischen Anthropologie*, Frankfurt: Suhrkamp.

Pouchelle, M-C. (1990) *The Body and Surgery in the Middle Ages*, New Brunswick, New Jersey: Rutgers University Press.

Robertson, R. and B.S. Turner (1989) 'Talcott Parsons and Modern Social Theory – an appreciation', *Theory, Culture & Society* vol. 6(4):539–58.

Rorty, R. (1986) 'Foucault and epistemology', pp. 41–50 in D.C. Hoy (ed.) *Foucault, a Critical Reader*, Oxford: Basil Blackwell.

Rosaldo, M.Z. and L. Lamphere (eds) (1973) *Woman, Culture and Society*, Stanford, California: Stanford University Press.

Rousselle, A. (1988) *Porneia, on Desire and the Body in Antiquity*, Oxford: Basil Blackwell.

Sacks, O. (1976) *Awakenings*, Harmondsworth: Penguin Books.

Sacks, O. (1981) *Migraine, Evolution of a Common Disorder*, London: Pan Books.

Sacks, O. (1984) *A Leg to Stand on*, London: Duckworth.

Schilder, P. (1923) *Das Körperschema*, Berlin: Springer.

Schilder, P. (1964) *The Image and Appearance of the Human Body*, New York: John Wiley.

Shapiro, K.J.S. (1985) *Bodily Reflective Modes, a Phenomenological Method for Psychology*, Durham: Duke University Press.

Sontag, S. (1988) *AIDS and its Metaphors*, New York: Farrar, Straus and Giroux.

Stauth, G. and B.S. Turner (1988) *Nietzsche's Dance, Resentment, Reciprocity and Resistance in Social Life*, Oxford: Basil Blackwell.

Suleiman, S.R. (ed.) (1986) *The Female Body in Western Culture, Contemporary Perspectives*, Cambridge, Mass: Harvard University Press.

Sydie, R.A. (1987) *Natural Women, Cultured Men, a Feminist Perspective on Sociological Theory*, Milton Keynes: Open University Press.

Turner, B.S. (1981) *For Weber, Essays in the Sociology of Fate*, London: Routledge & Kegan Paul.

Turner, B.S. (1982) 'The discourse of diet', *Theory, Culture & Society* vol. 1(1):23–32.

Turner, B.S. (1983) *Religion and Social Theory*, London: Heinemann.

Turner, B.S. (1984) *The Body and Society, Explorations in Social Theory*, Oxford: Basil Blackwell.

Turner, B.S. (1987) *Medical Power and Social Knowledge*, London: Sage.

Turner, B.S. (ed.) (1990) *Theories of Modernity and Postmodernity*, London: Sage.

Turner, B.S. (1991a) 'Missing bodies, towards a sociology of embodiment', *Sociology of Health and Illness* vol. 13(2):265–72.

Turner, B.S. (1991b) 'Recent developments in the theory of the body', pp. 1–35 in M. Featherstone *et al.* (eds) *The Body, Social Process and Cultural Theory*, London: Sage.

Urry, J. (1991) 'Time and space in Giddens' social theory', pp. 160–75 in C.G.A. Bryant and D. Jary (eds) *Giddens' Theory of Structuration, a Critical Appreciation*, London: Routledge.

Wallace, R.A. (ed.) (1989) *Feminism and Sociological Theory*, Newbury Park, Sage.

Weber, M. (1949) *The Methodology of the Social Sciences*, New York: The Free Press.

Weber, M. (1978) *Economy and Society*, Berkeley, Los Angeles and

London: California University Press.

Wilson, M.D. (1978) *Descartes*, London: Routledge & Kegan Paul.

Wolin, R. (1990) *The Politics of Being, the Political Thought of Martin Heidegger*, New York: Columbia University Press.

Zola, I.K. (1982) *Missing Pieces*, Philadelphia: Temple University Press.

Zola, I.K. (1991) 'Bringing our bodies and ourselves back in: reflections on a past, present and future "medical sociology"', *The Journal of Health and Social Behavior* vol. 32(1):1–16.

Chapter 2

The absent body in structuration theory

In the various approaches to the problem of agency and structure in sociology (which for the sake of convenience I shall refer to simply as 'structuration theory'), three issues have become predominant. The first is the nature of 'structure'; for example, is structure a set of rules, or is it a structure of objective constraints? The second issue has been to describe the nature of the agent in agency; for example, Anthony Giddens has strongly emphasized the knowledgeability of the agent from within a hermeneutic tradition. The third issue has been about the relationship between agency and structure; for example, can we transcend the traditional dichotomy between agency and structure in a theory of structuration? The first question raises problems in epistemology between idealism and materialism. The second issue relates to the Cartesian problem of mind and body. The third question can be taken as a version of the Hobbesian question of order. In short, the debate about the structure of social action encapsulates the core issues of modern sociological theory.

This discussion concentrates almost exclusively on the second of these issues, namely the Cartesian problem of embodied, conscious action, by attempting to develop a more satisfactory and comprehensive approach to the question: who or what is the 'agent' in structuration theory? The answer to the question is approached through an outline of a sociology of the body. The background assumption is that sociology, both classical and contemporary theory, has operated with a simplistic Cartesian dichotomy of mind and body, whereby 'body' is implicitly relegated to conditions of action. 'Body' is thus implicitly treated as a topic which falls within the field of the biological sciences and therefore outside the domain of sociology. While sociologists recognize that this dichotomy is not a satisfactory solution to problems raised within rationalist theories of action, until recently very few attempts have been made to incorporate the body in mainstream sociology.

The specific theme is that sociology has generally operated with an implicit (and occasionally explicit) rational model of disembodied man (the gendered noun is used deliberately) which has emphasized the voluntarism of the agent by an emphasis on choice and knowledge. What characterizes *homo sociologicus* is the capacity to make choices between various goals and means of action by reference to information and in terms of morality. The model against which classical sociology took its point of reference was classical and neo-classical economics. The problem of rationality in sociological theory was primarily formulated within the legacy of *economic* theories of action. Talcott Parsons's *The Structure of Social Action* (1937) provides a powerful critique of the economizing paradigms of action and social interaction by correctly emphasizing the importance of values and norms in social action, and by noting that the meaning of action from the point of view of the actor cannot be adequately integrated into a positivistic (in Parsons's terms) epistemology. Although rational choice theory is alive and well, sociology has typically adopted the position that action cannot be exclusively explained or under-stood by reference to utilitarian norms of maximization. But this critique of behaviouralism has unfortunately disembodied the agent who now appears as, to misquote Bentham, rationality on stilts.

Although these responses within (broadly) interpretative sociology are perfectly legitimate, they have not fully addressed the problem of the place of feeling and emotion in social action; this absence is a product of the neglect of a phenomenology of embodiment in classical sociology. Because the 'body' has been mistakenly treated as *only* a phenomenon occurring in the natural world, sociology has developed a 'blind spot' to certain universal existen-tial features of human action and interaction. Therefore, mainstream sociology has yet to address the emerging litera-ture on the sociology of the body (Frank 1991) in order to come to terms with some persistent, but often hidden, problems in the basic concepts of actor, action and action frame of reference. One associated issue in this absence of the body is that emotion-ality and affect have been forced into a residual category, because very little attention has been given to the embodiment of human agents.

There are, of course, a number of important exceptions to this rule, not all of whom could be considered in this discussion. However, it is important to note that an implicit theory of the body was essential to the symbolic interactionist tradition, to the sociological tradition represented by Erving Goffman, and also to the figurational approach of Norbert Elias. The nature of the body and the emotions are important, but implicit features of the theory of the civilizational process. In addition, in his symbol theory, Elias (1990) has defended the idea that symbols are also tangible sound-patterns which are developed as a consequence of the evolution of the human vocal apparatus. Elias thus connected the development of symbolization with the fact that human beings have a particular type of embodied vocal apparatus. However, Elias did not develop the metatheoretical presuppositions which stood behind his historical account of the transformation of court society (Elias 1987).

While sociology has at best developed a partial or implicit theory of the body in action theory, for various complex reasons, anthropology has developed an anthropology of the body (in the work of Mary Douglas, Marcel Mauss, Robert Hertz and many others). Anthropological studies have tended to concentrate on the symbolic significance of the body and on the use of the body in the determination of identity in sacrifice, mortification, and scarification in rites of passage, but the analytical implications of these anthropological studies for basic concepts of agent, agency and action in sociology have yet to be worked out.

The question of the body (and by association emotions, feelings and affect) has arisen in modern sociology for a great variety of reasons, but three developments are particularly important for this discussion: the postmodern critique of rationalist grand narratives, feminist critiques of patriarchal modes of analysis and the changing nature of consumerism, especially in terms of its consequences for aesthetics and ethics. Thus, it is evident that in general terms the growing interest in embodiment in sociology is an effect of complex social and intellectual changes. The postmodernist critique of traditional ideas of representation and thought within a classical rationalist model has brought the question of desire, sexuality and transgression more firmly into sociological debate. Social theorists like Jean Baudrillard (1975) have challenged the productivist metaphor in Marxist sociology to suggest alternative metaphors of sociability based on pleasures, waste and sacrifice. Here again the roots of this debate can be traced back through George Bataille,

Michel Foucault and Antonin Artaud to Nietzsche (Stauth and Turner 1988). On the other hand, feminist critiques of (male) views of the relationship between mind and body are also significant, especially in the work of J. Kristeva. Finally, there may also be important changes in the relationship between culture and structure in postindustrialism which have given a greater emphasis to the body, hedonism, emotionality and sensuality in both high and low cultures.

In order to pursue this issue, Talcott Parsons's theoretical strategy of the 'residual category' from his early critique of utilitarian positivism in *The Structure of Social Action* will be adopted. In the idea of the residual category, Parsons developed a powerful analytical strategy for locating and uncovering the theoretical weaknesses of various systems of thought. Residual categories are facts or observations which cannot be logically explained or accounted for in terms of the principal analytical components of a system of thought, and thus prevent the logical closedness of a system being fully developed. Parsons observed that

> every system, including both its theoretical propositions and its main relevant insights, may be visualized as an illuminated spot enveloped by darkness. The logical name for the darkness is, in general 'residual category'. Their role may be deduced from the inherent necessity of a system to become logically closed.
> (Parsons 1937:17)

Residual categories therefore prevent the development of a logically consistent and closed theoretical system of analysis, because the categories cannot be explained in terms of the main components of a theory. The system is closed by arbitrary, *ad hoc* devices. Thus, in order to explain social order, economic theory (based on rationalist and positivistic assumptions) had to resort to procedures which were not compatible with the foundations of economic rationality: these included 'moral sentiments' (Smith), 'habits and customs of the people' (Pareto), 'wants adjusted to activities' (Marshall) and 'instinct of workmanship' (Veblen). The analytical system thus remains in a state of crisis which threatens the logical structure of the entire system because it cannot achieve a satisfactory closure; its attempt at closure is always pre-emptive and premature. By uncovering these theoretical flaws in rational and positivistic theories of action and order, Parsons was able to develop his voluntaristic action framework which, through the various notions of internalization of values, autonomy of levels of explanation, the concept of the non-rational, the double contingency of action, and so forth, attempted to provide a systematic way of analysing hitherto residual categories in order to provide a simultaneous solution to the problems of social order and the voluntary character of social action.

At this stage I am less concerned with the validity of Parsons's action theory (Alexander 1984; Gould 1989); my aim is to use Parsons's analytical strategy of the 'residual category' to argue that mainstream sociology has (1) developed structuration theory with little or no reference to the embodiment of the agent; (2) however, sociologists constantly presuppose some notions about embodiment, but these assumptions are part of what Parsons called 'the darkness' of social theory; and (3) the darkness creates a problem in sociological theory which requires a solution. I propose that the solution is to be found in a sociology of the body which would allow us to develop a more satisfactory notion of who or what is the agent in structuration theory. This exercise is not marginal or irrelevant in social theory. The absence of a clear theory of embodiment 'disturbs' the fundamental conceptual apparatus of the sociology of action, for example in the famous division between behaviour and action. On the positive side, the development of a sociology of the body would have important implications for the *theoretical* development of medical sociology and the sociology of religion.

Thus, taking account of the embodiment of the human agent has important implications for four areas of contemporary sociology. First, it should bring about changes in the fundamental concepts of sociology, at least for sociology as a science of action. Second, it would provide a platform for working out the implications of contemporary critiques of Cartesian rationalism (especially in postmodernism and feminism), but it would also establish some linkage with earlier attempts (in philosophical anthropology and *Lebensphilosophie*) to work out a productive theoretical relationship between biology, phenomenology and sociology. Third, and very much in the spirit of Parsons's own action frame of reference, a sociology of the body offers some possibilities for interdisciplinary research and theory development between the natural and the social sciences. Finally, it offers a strong analytical grounding for a number of sub-disciplines inside sociology (especially medical sociology, sociology of gender, sociology of emotions and the sociology of food) which would reduce the fragmentation and specialization of various sub-branches of sociology. For example, the separation of medical sociology from mainstream theoretical development requires some theoretical repair (Gerhardt 1989; Turner 1987), which a sociology of the body could provide.

These arguments are considered initially through a commentary on the sociology of action as it was developed by Max Weber and Talcott Parsons. I shall examine the way in which the sociology of the body remained dormant in Peter L. Berger and Thomas Luckmann's work on the social construction of reality. The notions of practice and structuration in the sociology of Anthony Giddens and Pierre Bourdieu are compared with reference to the body. In conclusion, I defend the idea of the embodied actor against a number of possible objections. These authors are not selected in an arbitrary fashion. Weber and Parsons have dominated the 'agenda' for voluntaristic theories of action. Although various attempts have been made to resolve certain problems in Weberian rational action theory (Schutz 1970), the Weberian legacy – such as the division between rational, traditional and affectual action – still informs the core of sociological theory. Parsons's attempt to develop a minimal theory of voluntarism in the unit act is still the best defence of a sociological as against a utilitarian paradigm of action. Berger and Luckmann are important, because their attempt to construct a general theory of the social world is an important synthesis of the sociological tradition with the work of Arnold Gehlen and (using these terms loosely) the German tradition of philosophical anthropology. Finally, Giddens's theory of structuration and Bourdieu's theory of practice provide an important comparative perspective on contemporary approaches to action and structure. Whereas the body has not been seriously addressed by Giddens. Bourdieu's work on distinction, the logic of practice and the habitus has a clear relevance for any attempt to come to terms theoretically with a sociology of the body.

Thus, in Weber's sociology, the body is entirely absent from the conceptualization of agent and agency, whereas in Parsons the body hovers ambiguously between the agent and the conditions of action in the unit act. In the sociology of knowledge perspective of Berger, there was a missed opportunity to elaborate a phenomenology of embodied action which would be consistent with the idea of intentionality. In Giddens's structuration theory, despite the influence of phenomenology, hermeneutics and ethnomethodology, the body appears briefly as a constraining feature of the conditions of action. In emphasizing knowledgeability, Giddens has unwittingly returned to Parsons's formulation of the organic constraints of action. The body has returned to the environment as a constraint facing the actor. Finally, I show how Bourdieu, possibly because of the influence of anthropology, produces a theory of the body in his work on habitus, disposition and practice. However, because of Bourdieu's residual structuralism or because of his dissatisfaction with the existential legacy of Sartre (Bourdieu 1990), a phenomenology of the embodied actor has not been developed in relation to the ideas of habitus and practice.

Karl Marx is a rather obvious gap in this brief overview of contributions to the sociology of action. In order to deal in a satisfactory way with Marx, it would be necessary to trace the debate from Feuerbach's sensualism through Engels's version of historical materialism to the work of Timpanaro (Turner 1984). In summary, the dilemmas of structuration theory are all present in the debate between the supporters of the early versus the mature Marx. Either Marxism is based on a sensualist ontology of Man as producer who reproduces himself in transforming the world, or Marxism is a science of the fundamental laws of the mode of production, in which the subject of humanism is dismissed. Clearly there is a 'body' in the young Marx's account of the ontology of Man in which there is a dialectical process between humanization of nature and the naturalization of Man. In the *Economic and Philosophical Manuscripts* of 1844, Marx argued that 'Nature is man's *inorganic body* . . . Man *lives* on nature – means that nature is his *body*, with which he must remain in continuous interchange if he is not to die' (Sayer 1989:184). This tradition of Marxist ontology was partly reincorporated back into sociology via Berger's account of institutions, but with a very different philosophical purpose. The doctrine of Man as a practical, conscious species-being has been constantly reasserted whenever Marxists have turned against the structuralist legacy of Scientific Marxism.

WEBER ON PERSONALITY AND ACTION

Weber's definitions of action, social action and sociology in *Economy and Society* have provided a broad consensus among sociologists about the nature of their subject matter. For Weber, 'action' is human behaviour 'when and to the extent that the agent or agents see it as subjectively *meaningful*' and action is 'social', the meaning, which is 'intended by the agent or agents involves a relation to *another* person's behaviour' (Runciman 1978:7). It is well known that Weber's building blocks for sociological theory appear to attack any reification of sociological terms. Thus, action which involves meaning 'can properly be applied only to the behaviour of one or more *individual* persons' (Runciman 1978:16). While sociology has to take seriously the collective concepts of other social science disciplines, for the sociologist '*individual* human beings' are alone 'the intelligible performers of meaningful actions' (Runciman 1978:17). It is on this basis that Weber came to define sociology as a science which attempts to interpret the meaning of the actions of individuals in social relationships.

These definitions also sought to exclude certain phenomena from sociological study. Thus, Weber said that 'a collision between two cyclists is a mere occurrence, like a natural event' (Runciman 1978:26). It counts as social action once they engage in inter-action – shouting at each other, threatening to cause each other further damage and so forth. Furthermore, external behaviour towards material objects is not social; internal behaviour such as solitary prayer is not 'social' action. Although Weber recognized that the dividing line between social action and mere behaviour is hazy, in principal the division is around interaction between individuals taking regard for each other's actions and the presence of a meaning which is attached to behaviour. Thus, a crowd of tourists simultaneously raising their umbrellas in a shower are not engaged in social interaction.

On the basis of these distinctions, Weber went on to identify types of social action. Action may be rational in terms of selecting means which are appropriate to a given end (*zweckrational*) or an action may be rational in attempting to achieve some absolute value (*wertrational*). By contrast, affective action is 'the result of current emotional impulses and states of feeling' (Runciman 1978:28). As an illustration of acting affectively, we can consider a person 'who acts in such a way as to achieve immediate satisfaction of a need for revenge, pleasure, abandonment, blissful contemplation, or the release of emotional impulses' (Runciman 1978:28–9). Finally, traditional behaviour involves 'the expression of a settled custom' (Runciman 1978:28). Traditional and affective action lie on and often outside the boundary between 'consciously meaningful behaviour' (Runciman 1978:28) and merely reactive, non-social behaviour.

These features of Weber's sociology are well known and require no further elaboration here. Instead my intention is to draw out certain implications and criticisms of Weber's approach which have relevance for a sociology of the body. First, it is, as Weber himself implied, difficult to sustain the distinction between merely reactive behaviour which is like a set of 'natural events', internal behaviour and meaningful social action. In practice, it is difficult to locate a set of events which genuinely fall outside sociological interest, because they have no meaningful quality or because they do not involve interaction. The 'world' we inhabit is already shot through with symbolic and social significance; the conceptual division between the social and the natural is already a social division. Take solitary prayer. Surely the solitary meditative monk locked within the seclusion of the anchorite's cell is involved in a meaningful interaction with God? Surely the Muslim facing Mecca involved in regular daily prayer is not simply submitting to a 'person' but to The Person (and in a sense the only Person)? Is Hamlet haunted by the ghost of his father involved in social interaction? Is a shepherd interacting with a sheep dog, in which the communication of meaningful symbols (such as whistles and shouts) is an essential feature of the practical activity of directing a flock, involved in social action? In Weber's own terms, these forms of behaviour must count as meaningful social action, because they have all the requirements of his definition: they involve individuals, interaction and meaning. These examples therefore raise the question: what does Weber include and exclude under the phrase 'individual human beings'? By taking a common-sense meaning of 'individual person', Weber (possibly deliberately) excludes a range of interesting phenomena, such as interacting with ghosts or with God (Turner 1983).

Weber's definition involves him, therefore, if we want to unpick his version of 'the unit act', in a philosophical problem about what will count as an individual person, and secondly in a sociological problem about how social agents recognize other social agents. Being a competent social agent involves memory. I have to store up a minimal amount of information in order to be able to recognize other social agents and especially to recognize *particular* agents, such as distinguishing between my wife, my daughter and my mother. Clearly, memory is social; it is a complex process of mutual recognition and collective storage of social memory. In everyday practices, the body is a crucial feature of identification and social memory. Paul Connerton (1989) in *How Societies Remember* has drawn our attention to how various 'bodily practices' are important in collective ceremonial as a feature of how societies remember and reconstruct the past. In primitive societies, who or what people are is often inscribed on their bodies by means of scarification. In modern societies, the way in which we dress our bodies is still important to recognizing and defining persons.

These examples are, however, relatively trivial. In human societies, and in particular in societies which have a clear emphasis on individualism, the ability to recognize not just 'social agents in general' but particular named individuals is an essential pre-requisite of successful social interation; in rather obvious terms, it is for this reason that most conversations between strangers have to start with the question: 'who are you?' The answer to these questions of identity is normally a 'biography', involving names, dates and places. In all future interactions, being able to recognize *that* particular individual will depend on context and body, that is putting together context, biography and body. Even in advanced industrial societies, identity depends, not just on a passport number, car driver's licence, and an address, but on having the correct body which corresponds to these official marks; for example, fingerprints and photographs are an important feature of identity in modern societies. Where the correct body is of paramount importance, we may even tag bodies, as with bodies entering surgical wards, new-born babies, or convicts, who may have numbers or signs inscribed on their bodies. In our society, blood groupings, medical histories on AIDS, medical cards, and instructions about physical circumstances relating to heart condition and other ailments become increasingly important. Whole bureaucracies come to have a financial and legal interest in parts of our bodies.

These issues take us rather deeply into distinguishing between practices of individuation, individuality, the individual and individualism (Abercrombie *et al.* 1986). However, at this stage I want to argue, possibly along grounds which would be familiar to an ethnomethodologist, that for all practical purposes being a social agent involves having a body. This practical fact is an essential aspect to routine problems of misrepresentation, kidnapping, false identity, going missing, and so forth. These issues are either taken for granted or suppressed in Weber's account of the fundamental categories of social action.

Secondly, Weber gives a privileged position to rational (especially means–end rationality) action in his typology of social actions.

As we have seen, both affectual and traditional action lie either on or beyond the line between meaningful and reactive behaviour. The reason for this position is that it is impossible to understand Weber's theory of social action without taking into account his ethical view of personality. In order to pinpoint this argument, I shall draw on A.T. Kronman (1983), who argues that Weber's idea of personality (that is, a relatively coherent life-project involving a system of rational and ethical regulation) was shaped by a liberal Protestant value system. First, the world has no intrinsic meaning apart from the significance given to it by God. Second, human beings are invited to follow ethical prescriptions rather than absolute or fixed laws of nature. According to Kronman, Weber's epistemology is compatible with this type of theology. The fact-value distinction and the centrality of interpretation follow from (or are compatible with) this idea of the meaninglessness of reality without cultural interpretation. Behaviour has no social significance without meaning; this proposition necessarily places affect and tradition on the margins. In Weber's own philosophy, it is striving after value which separates man-with-personality from animal-with-affect. For Weber, personality must involve this inner-worldly idea of personal regulation to achieve mastery over the body as the seat of distracting emotionality (Holton and Turner 1989:83). Of course, Weber's ethical position is a good deal more complicated than this, because he remained ambiguous about the hedonistic enjoyment of life as against its rational control (Lepenies 1985). In his formal sociology, however, the problem of the body in relation to the 'individual human beings' whom he recognizes as the agents of social action was excluded. This analytical seclusion was an effect of his value system in which only 'personality' and 'rationality' have ethical significance.

PARSONS AND THE BIOLOGICAL CONDITIONS OF ACTION

Whereas in Weber it is the human individual who is the basic building block of the theory of action, Parsons (1937) started with the unit act, composed of an actor, conditions of action, normative standards and a goal. There are two crucial features of the unit act: the actor in a context of choice and values which cannot be reduced to conditions. The agent is a moral actor who is necessarily faced by certain dilemmas of action, such as the pattern variables.

The normative guidelines which are important for this moral being are derived from higher-order values which are in turn crucial for the stability of the unit act and ultimately for the entire social system.

In Parsons's action theory, the body remained a residual category, partly because, in rejecting biological reductionism, he came to allocate the body to the biological conditions of action. However, an implicit theory of the body remained important in his medical sociology, his views on affectivity, and in his analysis of the human condition (Holton and Turner 1986). When in his early work Parsons was preoccupied with the relationship between the cultural, social and personality systems, then the question of the biological roots of action (in the nature/culture distinction) was an issue which he could hardly ignore. Also Parsons's encounter with Freudian psychoanalysis also pushed him towards some consideration of the organic location of sexual drives. In the middle period of his academic career, his preoccupation with differentiation and integration, and with the various media of exchange, did not promote any primary concern with the sociology of the body, at least not with a debate over the biological conditions of human action. That is, while Parsons was thinking mainly about the development of research on American society – which was intended to become *The American Societal Community* (Parsons 1990) – he was engaged with the construction of an alternative view of American society, which did not involve any theoretical consideration of the relationship between the biological and the social. However, in the final phase of his intellectual career, the existential problems which faced Parsons in his late engagement with death and the meaning of life drove him more and more to engage with the dilemmas of human embodiment. In short, the implicit question of the human body assumed a number of different locations in Parsons's total intellectual production.

While rejecting biological reductionism, Parsons was primarily engaged in a debate with economic theory, and therefore with economic theories of action. Parsons's conception of the agent in the unit act was shaped by the legacy of economic theory in which an actor undertakes social actions in order to satisfy basic wants and needs. Of course, Parsons departed significantly from conventional views of wants and needs, because he recognized that the propensity to consume is shaped not in an economic context of scarcity but by values and norms in the cultural system. Nevertheless, the primary model of the unit act was economic. Parsons was also influenced from an early period by medical ideas and by the biological sciences. Walter Cannon's *The Wisdom of the Body* was influential for example in Parsons's views on functionalism. I would therefore argue that, while economic theory contributed to his development of a voluntary theory of action, biological research (in a broad sense) influenced the way in which Parsons came to think about social systems.

In *The Structure of Social Action* the fact that the human agent is also a living organism is conceptualized in terms of the environmental conditions of action, which to some extent limit and constrain the voluntary character of action. There is a tension between the normative character of choice (voluntarism) and the biological constraints of existence (determinism). In his later formulation of sociology in terms of personality, social system and culture, Parsons came to see the 'behavioral organism' as that feature of the agent which mediates between the environment and the cultural system. Action systems are located within a physical-organic environment which is mediated by the behavioural organism, and the general environment which is mediated by the symbols within the cultural system.

Within this cybernetic model, the organic conditions of life are in essence regulated at a lower level by the higher levels of culture. Within this paradigm of social control, Parsons came to make a sharper and more precise distinction between the human organism and the 'physical-organic environment'. For Parsons, the organic world was 'below' action in the hierarchy of elements or dimensions of a system of control. However, as human beings we can only know the physical world through the human organism, that is through the body, because our minds have no direct experience of external reality. We rely upon the brain to process information about this external context. Whereas, in *The Structure of Social Action*, the body was part of the biological and external conditions of action, in his middle period Parsons saw the human body as an organism mediating between mind and the physical environment, but in presenting the biological in that fashion he also implicitly accepted a definite mind–body distinction.

We can find another version of the interaction between the organic and the social system in Parsons's *The Social System* (1951) and *Toward a General Theory of Action* (Parsons and Shils 1951). In these two studies, Parsons introduced the idea of 'need-dispositions' which can only be satisfied through social interaction. The problem of need-dispositions is now fully articulated in terms of the two basic problems of the social system, namely allocation of resources and integration through values. These questions of socialization in relation to sexual gratification provided, for example, the background for Parsons's analyses of the incest taboo, the social role of women and the significance of the biologically related family unit. However, as Parsons became increasingly involved in the analysis of social systems in terms of differentiation and integration, the problem of the organic or biological conditions of action faded once more into the analytical background. The organic conditions of action became once again an environmental issue of the adaptive sub-system, relating mainly to the issue of economic facilities.

In his final intellectual stage of development prior to his death, Parsons in the 1970s produced a number of influential articles and essays which in general terms reflected upon the human condition, of which the biological was a significant aspect. These essays ranged over the question of 'the gift of life', the religious symbolism of life and death, the problem of the sick role (revisited), and theoretical questions concerning health and sickness from the point of view of action theory. Many of these contributions were eventually collected together in *Action Theory and the Human Condition* (Parsons 1978).

Thus, at various points in his intellectual development, Parsons touched upon the nature of the organic system in relation to action and system, without developing an articulate position on the nature of the body in relation to the actor. In part, Parsons recognized that the various levels of analysis were mere abstractions. He said 'there is no concrete human individual who is not an organism, a personality, a member of the social system, and a participant in a cultural system' (Parsons 1970:44). Nevertheless, his treatment of the behavioural system as the adaptive sub-system has been criticized for its lack of clarity and its underdevelopment in relation to the other sub-systems. For example, the conceptualization of the adaptive sub-system of personality has been criticized as 'the most poorly articulated statement of the theory of action standing on the comparable level of generality and importance in the body of theory as a whole' (Lidz and Lidz 1976:195). In order for Parsons to include the organism in a theory of *voluntaristic* action, it would be necessary to show how organic processes become significantly involved in meaningful action. Lidz and Lidz have argued that Parsons placed the behavioural organism too 'low' in the hierarchy of control; they attempted to adopt Piaget's psychology to 'up-grade', so to speak, the biological conditions of action within action theory. In refining the idea of the behavioural system, they divided the id into physiological needs and drives, the behavioural or intelligence system representing needs, and the emotional sector of personality (Lidz and Lidz 1976:218). Although I shall go on to argue for a phenomenological alternative to Piaget, Lidz and Lidz are perfectly correct in suggesting that the behavioural system has to be redefined to make it consistent with the action-theory assumptions of Parsonian sociology.

SOCIAL-CONSTRUCTION THEORY: THE ONTOLOGY OF ACTION

Perhaps the most sustained attempt in contemporary sociology to work out a systematic theoretical bridge between the voluntaristic theory of action and the biological foundations of the human organism (as a set of constraints) can be found in the work of Peter L. Berger and Thomas Luckmann. Berger's own work was thoroughly influenced by the contributions of Arnold Gehlen, whose *Man* (Gehlen 1989) was recently translated and published in English. Berger has been quite explicit in acknowledging Gehlen's general contribution to the theory of social institutions in Berger's own work (Berger and Kellner 1965). It was also spelt out in Berger's introduction to Gehlen's *Man in the Age of Technology* (1980). Gehlen's work was based fundamentally on the concept of man 'as a not-yet-finished being' in the philosophy of Nietzsche. Because human beings are not born into an environment to which they belong specifically, and because human beings are dependent for a very long period of maturation, the human world requires extensive institutionalization. Whereas other mammals are born already equipped to cope with a specific environment, humans are relatively flexible in terms of the range of environments into which they can be slotted. In Berger's terms, human beings (1) do not have a fixed set of instincts, the satisfaction of which is attached to a limited range of objects; and (2) humans do not have a species-specific environment. For example, human sexuality can find gratification in terms of a more or less infinite range of practices, partners and objects, and humans have successfully colonized every climate of the earth with the further option of occupying other planets in the near future. Gehlen's principal contribution, however, was not a sociology of the body as such but a sociology of institutions which were seen as a 'relief' from the uncertainty of man's diffuse instinctual constitution.

While there is this (so to speak) biological freedom, Berger and Luckmann argued in *The Social Construction of Reality* (1966) that there are correspondingly cultural and social constraints on humans. In the absence of a species-specific reality, human societies are involved in the endless tasks of social construction of social orders, the production of social legitimation and the maintenance of 'plausibility' structures. Against the infinite possibilities of social construction, human beings are faced by the threat of chaos, de-legitimization and consequently homelessness. It is of some interest that both Berger (1967) and Luckmann (1957) found in the study of religion the most potent example of their general argument.

They conceptualized the social in terms of three moments as a summary of the structuration problem in classical sociology. These three moments are externalization, objectification and internalization: humans make the social reality in which they live; this social reality in turn shapes human experience of reality; and the social reality is an objective structure which determines social agents. In *The Sacred Canopy* Berger (1967:4) comments that 'It is through externalization that society is a human product. It is through objectivation that society becomes a reality sui generis. It is through internalization that man is a product of society'. These propositions can be seen as a summary of the theories of alienation and anomie, of Marx's claim in *The 18th Brumaire of Louis Bonaparte* that 'Men make their own history, but they do not make it just as they please' and of the theory of the *conscience collective* in Durkheim. Berger, however, gave a rather interesting twist to the Marxist theory of reification as alienation. Although anomie is a threat to the stability of the plausibility structures which support the cultural edifice of human societies, human beings require reification, because they need to experience the subjectively precarious world as objectively valid. It is only through these processes that the necessary facticity of everyday reality can be guaranteed. It is only when the frailty of social relations can be experienced as an external, objective and reliable fact that humans can have confidence in the security and validity of the world. The satisfaction of the problem of theodicy requires the alienation of human beings.

In religion, these processes of objectification are seen in their most pristine form. With some degree of irony, Berger agreed with Marx's view that religion is alienation, that is the human spirit externalized in social relations which are then experienced as alien to human practice. However, Berger argued that humans require the alienation of the social environment through religious consciousness, if they are to protect themselves from the potential chaos of meaninglessness and anomie. In religion, the plausibility of belief – the classical problem in theodicy of explaining a just god in an unjust world (Turner 1981) – depends on the continuity of 'the sacred canopy', and in the modern world this stability is gradually undermined by the pluralization of life-worlds. The traditional religious order is gradually converted into a supermarket of pluralistic beliefs. The gradual de-institutionalization of social reality means that we are all 'homeless' (Berger *et al.* 1973).

Berger and Luckmann have produced a general theory of structuration which certainly takes the body seriously. As we have seen, the theory attempts to combine philosophical anthropology and sociology in which certain features of social life – the need to legitimize social arrangements – are derived from man's unfinished nature. In evaluating this theory, I shall concentrate exclusively on the work of Peter Berger – the collaboration with Luckmann was significant but limited. Having considered some general criticisms of Berger's sociology, I then make some specific remarks on the role of the body in his theory of structuration. Probably the most damaging criticism of his work is that he has conflated and confused certainty and meaning (Abercrombie 1986). The ontologically unfinished nature of human beings may biologically predispose them to a need for order, stability and certainty; it may also be the case that human beings require their world to be more or less meaningful, but these are two separate and distinctive requirements. Human beings may find their world very uncertain, while also finding it meaningful. Alternatively, human beings may find their world stable and predictable, but not necessarily meaningful. Berger has combined the need for meaning (at the level of the individual) with the need for order (at the level of the social system), because his sociology ultimately precludes agency in favour of structure. Given the ever-present terror of chaos, people are committed to order, because they are psychologically unable to live with the possibility of meaninglessness. Berger argues that 'Men are congenitally compelled to impose a meaningful order on reality' (Berger 1967:22). The consequence of collapsing order and meaning into a single category is two-fold: first it brings God back into sociology via the back door, and second, it precludes any possibility of the autonomy of the agent (Abercrombie 1986:29).

What then is the role of the body in Berger's sociology? The body enters Berger's sociology via phenomenology. Following both Merleau-Ponty and Schutz, Berger is concerned to understand our intentionality towards the life-world, of which the body is a fundamental feature. Berger has been interested throughout his academic career in the dialectical tensions between self and body, self and society. He expresses this dialectic in the idea that 'man *is* a body, in the same way that this may be said of every other animal organism. On the other hand, man *has* a body. That is, man experiences himself as an entity that is not identical with his body, but that, on the contrary, has that body at its disposal' (Berger and Luckmann 1966:48). Although the embodiment of the agent is a condition of constraint, human beings have an intentionality by which they experience the body which is at their disposal. Although Berger's theory of the body in agency is thus fully grounded in the work of Gehlen and Husserl, he did not develop the basic idea of the dialectic between having and being a body in terms of everyday reality. For example, human embodiment does not play a prominent place in Berger's account of identity (1966). In his article on 'Identity as a problem in the sociology of knowledge', Berger in footnote seventeen in connection with the relationship between cognition and physiological processes refers to certain 'intriguing possibilities' for a 'socio-somantics' in Georg Simmel's discussion of the possibility of a sociology of the senses in his *Soziologie* and in Marcel Mauss's essay on 'techniques of the body', but he did not attempt to integrate the ideas of intentionality, phenomenology of the body and the social dynamics of identity. Berger did not develop any elaborate view on how the body is connected to the experiences of others in the formation of the self, of how the ageing of the body changes social identities, or of how traumatic damage to the body (such as the loss of a limb) changes the experience of and intentionality towards the world.

How may we explain this gap? Stephen Ainlay (1986) speculates that, because Berger has drawn his phenomenological inspiration primarily from Schutz in whose work the body remained an aside in the study of the other, language and the idea of projects 'within reach', the body also remained entirely peripheral to Berger's main preoccupation with the servicing of the structures that make reality plausible. Of course, Schutz had in his early work – such as

Der sinnhafte Aufbau der sozialen Welt (1932) – taken notice of the idea of the 'unity of the body', the somatic experience of being 'here', and thus the body was important in the discussion of temporality. For Schutz, this knowledge of finiteness (of the body) was an important condition in ascertaining the meaning of human existence. All human being is being-for-death. Nevertheless, Schutz's treatment of the body was underexplored. Had Berger derived his phenomenology from Merleau-Ponty, Sartre and Marcel, then the theoretical outcome would have been rather different. Having offered some explanation for the underdevelopment of this side of Berger's sociology, Ainlay concludes by asserting, correctly in my view, that 'the need to integrate a greater appreciation of the body into sociology – especially interpretive sociology – is glaringly apparent. Developing an adequate sociology of the body is certainly among the chief mandates for a phenomenologically inspired interpretive sociology' (Ainlay 1986:52).

Although this lack of engagement with French phenomenology may have some bearing on the problem, it is difficult to see how a sociologist with a special interest in religion could avoid or miss the importance of the body in the fundamental rituals of social life, especially in so-called primitive societies. The whole question of pollution, sacrifice, purification and rites of passage is so fundamental to the relationship between religious culture, social process and the body that it is difficult to imagine how this issue has been so systematically neglected, not simply by Berger of course but by mainstream sociology of religion as a whole.

TWO MODERN VERSIONS OF STRUCTURATION: GIDDENS AND BOURDIEU

Giddens's treatment of the agency-structure issue has dominated much of his mature sociology. This theory has recently been the topic of much critical assessment (Bryant and Jary 1990; Clark 1990; Cohen 1989; Held and Thompson 1990) and I do not intend to review all aspects of his contribution. However, it is possible to summarize some of the principal features of Giddens's version of structuration. First, Giddens has tried to defend the idea of the knowledgeability of the social actor, that is, actors are not puppets of objective structures and they are not dupes of cultural systems. While Giddens recognizes that not all actions are purposeful, in principle the knowledgeable agent is able, in fact must be able, to monitor his/her actions reflexively. Similarly, Giddens recognizes the presence of the unconscious in human actions, but he has adopted a rather different terminology of the unconscious, practical consciousness and discursive consciousness. Furthermore, the importance of the unintended consequences of action is a significant feature of his general theory, but these unintended consequences, along with unacknowledged conditions of action, are a significant feature of his views on structure.

Secondly, he has attempted to avoid the static dualism of agency and structure. By contrast, Giddens insists that we must examine the dynamic dialectic of social reproduction, that is how in everyday life the social structures, which make action at all possible, are reproduced in the very performance of those actions. Thus, Giddens has been increasingly concerned with the temporal features of action, with the 'duality of action' and with the 'recursive character' of social life (Giddens 1976; 1979; 1984). For Giddens, 'structure' is not the relations of interaction which constitute social structures; structure refers to the systems of generative rules and resources in society, and in turn these rules and resources are 'properties of social systems'.

The point of both contributions to structuration theory is in practice to take the idea of sociology as an interpretative approach to action seriously, but at the same time to overcome the conventional dichotomies of action and structure, objective and subjective, materialist and idealist epistemologies which have characterized much of the modern history of theoretical sociology. By taking action seriously, Giddens has learnt the lessons of Schutz's critique of the Parsonian action scheme (Grathof 1978), Garfinkel's transformation (1967) of the Parsonian analysis of order, of the Wittgensteinian notion of rules and games, and finally of Goffman's ethnographies of self and place. In short, Giddens has attempted a brilliant synthesis of the various strands of action theory in order primarily to overcome both functionalism and historical materialism insofar as they preclude the knowledgeable actor.

The principal criticisms of Giddens are that: (1) his account of 'structure', depending on the vague notion of 'rule' and enabling conditions, has in effect excluded any idea of structure as constraint (Thompson 1990); (2) the idea of the duality of structure in fact involves a vacuous notion of interaction which does not in fact overcome dualism (Smith and Turner 1986); (3) the synthesis of hermeneutics, action theories and Wittgensteinian philosophy has not in fact produced an original paradigm; and the same argument holds for Giddens's use of geography in the analysis of space; and (4) structuration is a description of the elements of action and structure which has little or no bearing on how sociologists conduct research. To this catalogue, I want to add the argument that Giddens has yet to address the question of the body of the agent in structuration theory, despite the few hints about embodiment which occur in *The Constitution of Society*, whenever Giddens turns to the theories of Goffman, or where he refers to the body as one of the constraints on action. In his recent study of *Modernity and Self-Identity*, Giddens (1991) has taken some notice of the fact that in modernity the body is a topic of reflexive inspection. He claims that the body is an important feature of the modern life-style. These observations have yet to be systematically built into structuration theory.

This absence of any serious analysis of embodiment is interesting given Giddens's obvious exposure to some features of the phenomenological tradition; it is also problematic given Giddens's declared interest in the question of gender (Giddens 1990:282-7); and finally it is curious given Giddens's obvious sympathy for the work of Erving Goffman (Giddens 1987:109-39). How does the body appear then in the work of Giddens? Primarily the body is treated as a constraint on action and therefore in some sense 'outside' the actor. Thus, in defending himself against the criticisms of John B. Thompson, Giddens (1990) referred back to *The Constitution of Society* (Giddens 1984:174-9) in which he had distinguished between three types of constraint, namely physical, power and structural constraints. Thus, 'there are physical constraints upon activity, deriving from the human body and the material environment' (Giddens 1990:258).

Because Giddens has, like Parsons in *The Structure of Social Action*, been primarily concerned to defend the idea of the knowledgeable, purposeful, active actor, he has had a built-in tendency to oppose any and all forms of biologism, organicism and so forth as versions of a positivistic reductionism. Hence, the body is relegated to a feature of the constraining environment; the body cannot enter action theory or structuration theory as a necessary feature of the agent and agency. The actor is essentially a thinking and choosing agent, not a feeling and being agent. There is an inconsistency here, since Giddens wants to draw on Goffman's idea of the self and display, but without reference to Goffman's implicit theory of the body. In Goffman's sociology, face-work, and hence the social body, is one feature of his approach to the representational self which is an embodied feature of social interaction. In this sense at least, Giddens turns the clock back to Parsons's solution in the unit act where the body, having been relegated to the sub-stratum of action, cannot feature as an essential aspect of perception, identity and practice in action.

BOURDIEU: THE BODY IN THE LOGIC OF PRACTICE

There are some parallels between Giddens and Bourdieu. Both theorists have developed a view of praxis which has been forged in opposition to the legacy of various types of functionalism and structuralism. Thus, Bourdieu's early anthropological research on the Kabyle produced a reaction against the structuralism of Lévi-Strauss, and gave rise eventually to his early theory of practice in *Outline of a Theory of Practice* (1977). Both writers have an ambiguous relationship to Marxism (Bourdieu 1990). In Giddens's case, we have the two critiques of historical materialism (Giddens 1981; 1985). In Bourdieu's sociology, the contribution of Marx to his work on cultural capital and social reproduction is explicit, and yet he has constantly challenged the spatial metaphors of determinism in Marxism (such as the base/superstructure model). Bourdieu's theory of practice is designed, like Giddens's structuration theory, to overcome these conventional oppositions. In Bourdieu's research the idea of practice is aimed at transcending any rigid dualism between 'mental structures' and 'the world of material objects'. Both sociologists are engaged with reflexivity, especially because the sociology of sociology is not an optional feature of doing sociology.

The body has a very general role in Bourdieu's sociology. Thus, the human body is crucial to his ideas about physical capital, which he occasionally subsumes under the broader category of cultural capital. However, the body as a social product under the social logic of physical capital is produced in a particular habitus by sport, leisure and consumerism. The body in this sense is the consequence of (class) practices. For example, weight lifting articulates working-class bodies, while jogging and tennis produce a body which is at ease in the middle-class milieu or habitus. Class practices are inscribed on the body, which is also a social product of special class activities. In Bourdieu's work, the human body is an essential component of the reproduction of class inequalities. The manifold ways in which social groups manage their bodies (through sport, diet, surgery, education and sexual practices) 'reveals the deepest dispositions of the habitus' (Bourdieu 1984:190).

The appeal of Bourdieu's work in this context is therefore perhaps obvious. Here is a *general* theory of practice, which attempts to resolve the classical subjective/objective, agency/structure dichotomies of social science by combining the idea of struggle over resources with a structurally organized habitus in terms of a theory of social practice. The scope of Bourdieu's work has only recently become generally recognized.

Although mainstream sociology outside France is probably very familiar with his work on cultural reproduction in *Reproduction in Education, Society and Culture* (Bourdieu and Passeron 1977), the full importance of *Le sens pratique* (1980), *Homo Academicus* (1988) or *Choses Dites* (1987) and of his many occasional essays and interviews has yet to be digested.

Apart from the scope of Bourdieu's work, it is important for its attempt to avoid any reduction of cultural to economic capital, and yet he retains a dominant sense of the way in which cultural objects are determined by social (primarily class) struggles. We can thus see society as an organization of fields which are the sites of individual and collective struggle over the production and consumption of cultural goods. The value of a symbolic good is determined by the quantity of symbolic capital which a producer has accumulated. Victory in these confrontations permits a dominant social group to exercise symbolic violence over other consumers in this cultural field.

Bourdieu's work is provocative because he attempts to show how in the world of high culture and within the academy, similar struggles for symbolic dominance are undertaken. Thus, it is important to recognize that *Distinction* is a critique of Kant's aesthetic theory in the *Critique of Aesthetic Judgement* in which Kant attempted to show that taste is disinterested, purposive and universally valid. By contrast, Bourdieu argues that taste is social, active and committed rather than disinterested. Taste, rather than universal, is specific to the habitus of a class. In fact Kantian taste encapsulates a disgust for vulgarity which seeks immediate enjoyment. In developing a *social* theory of taste against Kant, I want to argue that Bourdieu ends up with a Durkheimian sociology of knowledge (or better a sociology of classification) which tends to undermine the emphasis on agency and struggle. There appears to be little room in Bourdieu's theory for successful struggle against the dominant classificatory scheme. It might appear that the highly structured and perfectly organized world of consumption in *Distinction* adequately describes the French field, but I want to claim that this disguised structuralism is the legacy of Bourdieu's location in French structuralism. In fact, in *Outline of a Theory of Practice*, he often adopts a clearly deterministic view of the habitus: 'As an acquired system of generative schemes objectively adjusted to the particular conditions in which it is constituted, the habitus engenders all the thoughts, all the perceptions, and all the actions consistent with those conditions, and no others' (Bourdieu 1977:95). It would not be surprising if an unsuspecting reader mistook these words for Durkheim's description of social facts. Indeed, it is perhaps no accident that *Outline of a Theory of Practice* was an attempt to generalize his anthropological work on Kabyle culture, and that Durkheim based his conception of mechanical solidarity in *The Division of Labour in Society* on contemporary studies of Kabyle segmentary social structure; Durkheim's own sociology of knowledge was also a reply to Kant, just as Bourdieu's theory of taste and disposition is a reply to Kantian aesthetics.

There are also important differences therefore between Giddens and Bourdieu, which reflect their different social and philosophical backgrounds; Giddens has been engaged with the legacy of classical sociology as mediated through Parsons, whereas Bourdieu is attempting to transform various traditions of French structuralism. One important difference is that, if Giddens has understated the nature of objective constraints in the structuring of action, then Bourdieu has understated the autonomy of action in his structured account of the habitus.

Another difference is around the question of the body. Bourdieu, as I have already suggested, requires a sociology of the body as part of his more general ideas about habitus and practice. In *Distinction*, as I have indicated, the representation of the body and the dispositions of the body are an important feature of physical and cultural capital. The body in Bourdieu appears as a site on which is inscribed the cultural practices of the various classes. Each class and each class fraction has, for example, a characteristic sport which exhibits both their economic and their cultural capital. Furthermore, in sport learning is best conveyed by illustration, because it is often impossible to communicate the practices which are necessary without bodies. Dance in particular can only be adequately conveyed visually. Training for sport thus presents Bourdieu with a powerful illustration of the relationship between social membership, regulation and embodiment. Thus, 'Bodily discipline is the instrument *par excellence* of every kind of "domestication"': it is well known how the pedagogy of the Jesuits made use of dancing ... The gesture ... reinforces the feeling which reinforces the gesture. Thus is explained the place that all totalitarian regimes give to collective bodily practices' (Bourdieu 1990:167). In more abstract terms, the body in Bourdieu's theory is a carrier of dispositions which are themselves the conduits of interests within the habitus or life-world of the actors. Although the habitus is a practical logic (and is therefore vague and indeterminant), Bourdieu writes about the habitus in a very deterministic fashion. The habitus is a system of dispositions with reference to a given place, which produces the regularities in modes of behaviour. In short, Bourdieu's theory appears to retain a deterministic and structuralist logic in which the body is primarily the bearer (*Träger* to use Weberian terminology) of cultural codes; there is little room in Bourdieu's work for some phenomenological understanding of the 'lived-body' as an essential aspect of action, intention and disposition.

THE BODY AND SOCIOLOGICAL THEORY

So far I have offered little more than a history of ideas which has examined the body as a residual category in the development of sociological theory, as illustrated in the work of Weber, Parsons, Berger, Giddens and Bourdieu. I want next to offer a more positive account of the importance of the body for sociological theory and to show how such a development might influence sociology as a practice. I shall then attempt to defend the argument against possible criticisms.

If we take the idea of sociology as a science of action, we need a theory of the body, because human agency involves more than mere knowledgeability, consciousness and intention. Of course, there are various types of 'agent' in sociology. As Margaret Archer (1988) has argued we should avoid the traditional conflation of the 'people' with the 'parts' by being clear about the difference between social system analysis and social analysis. If 'collective' agency refers to such entities as class and state, then the question of body may be irrelevant – apart, that is, from the persistence of metaphorical notions of 'the body politic'. If, however, we are interested in human agents at the social rather than the system level, then it is difficult to see how we can avoid a sociology of the body.

By virtue of being embodied, human agents are subject to certain common processes which, while having biological, physiological and generally organic foundations, are necessarily social. These common processes are related to conception, gestation, birth, maturation, death and disintegration. Since so many social practices are based on such obvious events – marriage, legitimate copulation, socialization, burial and rituals of grief – it is strange that sociology has so generally neglected these practices as evidence of our embodiment. In concentrating on the meaningfulness of social action from the point of view of the actor, sociology has avoided this corporeal side of action, despite the fact that questions of meaning (Weber's theodicy problem) are almost invariably associated with our embodiment – suffering, pain, joy and death, to take some rather banal illustrations. While *homo sociologicus* is busy choosing goals in terms of ends to achieve certain purposes, he/she is apparently liberated from the body, despite the fact that in 'real life' body-maintenance practices occupy most of our daily routines. Body practices – cleaning, washing, grooming, presenting, disciplining, disguising, stimulating – dominate everyday *social* life, although they do not appear in sociology.

This aspect of human agency is not somehow beyond or even alongside the social; the body and human embodiment are social. To insist on the importance of a phenomenology of the body is thus to deny that 'behaviour' is an area of analysis which is not sociological. My relationship to my body and to other bodies is social. For reasons outlined by Marcel Mauss and others, even our basic 'body techniques' such as walking, standing or sitting are 'social constructions'; they are developed modes of operation which are specific to given cultures, but it is also the case that I develop 'my' walk. More fundamentally, perception, for reasons again outlined by Merleau-Ponty and others, is intimately bound up with my embodiment, or with what Schutz had in mind by the notion of relevance. Perceptions of 'above', 'below', 'beyond' and so forth cannot be separated from, or indeed made sense of, without reference to my body in relation to other (social) bodies. Having a pain is necessarily a social event, because I need a language by which to work out notions of 'I feel a pain in my leg'. It is difficult to see how a sociology of action can get along without a sociology of the body, because any sociology of action must make some presuppositions about embodied agency, sentiments and emotions, and perceptions.

My second argument, which was rehearsed earlier in this paper, is that most existing theories of action make assumptions about the identity of agents without actually questioning or elaborating those assumptions. All social action makes assumptions about identifying 'other social actors'. Weber, as we have seen, made assumptions about God not being a social actor in his illustration of solitary prayer. What will count as 'an actor' will depend on certain cultural codes; for example, in some societies like traditional Bali ghosts or spirits are an important part of the social scene. My identity, however, depends in part on my occupancy of a body, or rather a particular body, namely my body. This claim is not to argue that there is a biological conditioning of identity, since my body is part of a complex social context involving many other relevant (social) bodies.

In short, a theory of social action has to make assumptions about feelings, emotions, perceptions, identities and the continuity of agents across space and time which only a sociological theory of the body can satisfy. The reasons for this neglect go deep into Western culture. Calvinism would be one reason for the residual nature of human embodiment in Western philosophy. Both Calvinism and classical philosophy tended to argue that the life of the spirit starts where the life of the body comes to an end. Classical sociology has argued that the life of the knowledgeable agent starts where the physical (that is the bodily) conditions of existence come to an end. It was for this reason that Horkheimer and Adorno in *Dialectic of the Enlightenment* (1973) referred to the secret history of Western culture as the suppressed history of the body.

There are a number of possible objections to the argument which I have presented. Let us consider three of them: (1) my argument forces me into a methodological individualism which most versions of structuration theory have attempted to avoid; (2) my argument has failed to clarify what would be meant by 'the body' – a clarification which might itself specify the relationship between agency and structure; and (3) no significant consequences follow from incorporating a sociology of the body into mainstream sociological theory. I shall not consider the objection that I am pushing sociology towards sociobiology; the argument from Gehlen and others emphasizes the unique biological features of human existence which makes any comparison with animals either trivial or pointless. I shall attempt briefly as a form of conclusion to address these difficulties.

CONCLUSION: INDIVIDUAL BODIES/SOCIAL POPULATIONS

In those phenomenological accounts of the body which have captured the imagination of social philosophers, it is often the individual experiences of people towards their bodies which provide the primary illustrations of embodiment. I have in mind the work of Oliver Sacks, Paul Schilder, Merleau-Ponty and even Freud himself. Does this point to some rather obvious problem that, since I have an individual body, the question of the body should be left to psychologists, to existential philosophers or to the natural sciences? Would a sociology of the body necessarily involve some assumptions about methodological individualism? I want to deny this possibility in order to defend the position as a whole.

One important argument for 'bringing the body back into sociology' (Frank 1990) is that it is difficult to understand how action would be possible at all without some capacity for individuation and identity. The body is the most obvious interactional and interpersonal carrier of such identities, but these identities are not individualistic. They rest on social recognition; they are based on collectively shared memories of individuals. The body is an essential feature of the storing of such memories over time. Hence the body is a social phenomenon, and is essential to the organizing of social phenomena.

This issue then raises the question: what is the body? There are basically two contrasted positions. In much contemporary French social theory, such as the work of Michel Foucault, the body is an effect of discourses. The rise of the docile body in *Discipline and Punish* (Foucault 1979) illustrates the idea that the body is the historical outcome of formations of power/knowledge which actually produce different orders of body. In other words, there is a structuralist legacy in Foucault's poststructuralism. By contrast, there is an alternative legacy of German philosophical anthropology, French existentialism and general phenomenology in which it is the 'lived body' and the embodiment of the actor which is the focus of investigation. This tradition embraced such diverse figures as Gehlen, Plessner, Sartre, Marcel, Merleau-Ponty and Schilder.

These writers had many different intellectual aims, but two are worth noting. First, there was an attempt to integrate or reconcile natural and social science approaches to organic life. Second, it involved a critique of a rationalist tendency to understand perception, conception and action without reference to the lived-body experience.

These two traditions – structuralist and phenomenologist – need not face us necessarily with an exclusionary theoretical choice. Obviously Foucault's work on the body and truth is extremely important, but it asks questions which are very different from those posed by Merleau-Ponty. A sociology of the body in fact reflects the two issues raised in Berger's sociology, namely the difference between having a body (the body as constraint) and being a body (the body as capacity). There is a duality in the body which is partly reflected in two contrasted philosophical traditions.

Finally, I have argued that the sociology of the body is a non-trivial corrective to mainstream sociological theory, which has important consequences for sociology. I shall mention only one. The theme of the body is an important corrective to the fragmentary and fissiparous character of sociology. The sociology of the body provides a platform for (re)integrating medical sociology back into mainstream sociological theory, while also providing a connecting theme or linkage between the sociology of ageing, the sociology of the emotions, and broadly feminist theory. The sociology of the body will come to have fundamental connections with postmodernism and with postmodern sociology, because the body has an uncertain and problematic status in the unfolding culture of postmodernity. For example, the interpenetration of the technological, biological and social worlds has given rise to a new entity (the cyborg) which cross-cuts the organic and the inorganic (Levidow and Robins 1989). The sociology of the body also provides a field or theoretical site where what we might call sociological ecologism could find connecting points with classical sociology. The body – like religion – raises the question which is finally the elementary question of sociology itself – what is the social?

REFERENCES

Abercrombie, N. (1986) 'Knowledge, order and human autonomy', pp. 11–30 in J.D. Hunter and S.C. Ainlay (eds) *Making Sense of Modern*

Times, Peter L. Berger and the Vision of Interpretive Sociology, London: Routledge & Kegan Paul.

Abercrombie, N., S. Hill and B.S. Turner (1986) *Sovereign Individuals of Capitalism*, London: Unwin Hyman.

Adorno, T.W. and M. Horkheimer (1973) *Dialectic of Enlightenment*, London: Routledge & Kegan Paul.

Ainlay, S.C. (1986) 'The encounter with phenomenology', pp. 31–56 in J.D. Hunter and S.C. Ainlay (eds) *Making Sense of Modern Times, Peter L. Berger and the Vision of Interpretive Sociology*, London: Routledge & Kegan Paul.

Alexander, J.C. (1984) *Theoretical Logic in Sociology. The Modern Reconstruction of Classical Thought: Talcott Parsons*, London: Routledge & Kegan Paul.

Archer, M. (1988) *Culture and Agency, the Place of Culture in Social Theory*, Cambridge: Cambridge University Press.

Baudrillard, J. (1975) *The Mirror of Production*, St Louis: Telos.

Bell, D. (1976) *The Cultural Contradictions of Capitalism*, New York: Basic Books.

Berger, P.L. (1966) 'Identity as a problem in the sociology of knowledge', *European Journal of Sociology* vol. 7:105–15.

Berger, P.L. (1967) *The Sacred Canopy*, Garden City, NY: Doubleday.

Berger, P.L., B. Berger and H. Kellner (1973) *The Homeless Mind*, New York: Random House.

Berger, P.L. and H. Kellner (1965) 'Arnold Gehlen and the theory of institutions', *Social Research* vol. 32:110–15.

Berger, P.L. and T. Luckmann (1966) *The Social Construction of Reality*, Garden City, NY: Doubleday.

Bourdieu, P. (1977) *Outline of a Theory of Practice*, Cambridge: Cambridge University Press.

Bourdieu, P. (1980) *Le sens pratique*, Paris: Editions de Minuit.

Bourdieu, P. (1984) *Distinction, a Social Critique of the Judgement of Taste*, London: Routledge & Kegan Paul.

Bourdieu, P. (1987) *Choses Dites*, Paris: Editions de Minuit.

Bourdieu, P. (1988) *Homo Academicus*, Stanford: Stanford University Press.

Bourdieu, P. (1990) *In Other Words, Essays Towards a Reflexive Sociology*, Cambridge: Polity Press.

Bourdieu P. and J-C. Passeron (1977) *Reproduction in Education, Society and Culture*, Newbury Park: Sage.

Bryant, C.G.A. and D. Jary (eds) (1990) *Giddens' Theory of Structuration, a Critical Appreciation*, London: Routledge.

Clark, J. (ed.) (1990) *Anthony Giddens, Consensus and Controversy*, Brighton: Falmer Press.

Cohen, I. (1989) *Structuration Theory*, Cambridge: Polity Press.

Connerton, P. (1989) *How Societies Remember*, Cambridge: Cambridge University Press.

Elias, N. (1987) *The Court Society*, Oxford: Basil Blackwell.

Elias, N. (1990) *The Symbol Theory*, London: Sage.

Foucault, M. (1979) *Discipline and Punish*, London: Tavistock.

Featherstone, M., M. Hepworth and B.S. Turner (eds) (1990) *The Body, Social Process and Cultural Theory*, London: Sage.

Frank, A. (1990) 'Bringing bodies back in', *Theory, Culture & Society* vol. 7(1):131–62.

Frank, A. (1991) 'For a sociology of the body: an analytical review', pp. 36–102 in M. Featherstone, M. Hepworth and B.S. Turner (eds) *The Body, Social Process and Cultural Theory*, London: Sage.

Garfinkel, H. (1967) *Studies in Ethnomethodology*, Englewood Cliffs, New Jersey: Prentice-Hall.

Gehlen, A. (1980) *Man in the Age of Technology*, New York: Columbia University Press.

Gehlen, A. (1989) *Man*, New York: Columbia University Press.

Gerhardt, U. (1989) *Ideas about Illness, an Intellectual and Political History of Medical Sociology*, London: Macmillan.

Giddens, A. (1976) *New Rules of Sociological Method, a Positive Critique of Interpretative Sociology*, London: Hutchinson.

Giddens, A. (1979) *Central Problems in Social Theory, Action, Structure and Contradiction in Social Analysis*, London: Macmillan.

Giddens, A. (1981) *A Contemporary Critique of Historical Materialism; Volume 1: Power, Property and the State*, London: Macmillan.

Giddens, A. (1984) *The Constitution of Society, Outline of a Theory of Structuration*, Cambridge: Polity Press.

Giddens, A. (1985) *The Nation-State and Violence.* Volume 2 of *A Contemporary Critique of Historical Materialism*, Cambridge: Polity Press.

Giddens, A. (1987) *Social Theory and Modern Sociology*, Cambridge: Polity Press.

Giddens, A. (1990) 'A reply to my critics', pp. 249–301 in D. Held and J.B. Thompson (eds) *Social Theory of Modern Societies*, Cambridge: Cambridge University Press.

Giddens, A. (1991) *Modernity and Self-Identity: Self and Society in the Late Modern Age*, Cambridge: Polity Press.

Gould, M. (1989) 'Voluntarism versus Utilitarianism: a critique of Camic's history of ideas', *Theory, Culture & Society* vol. 6(4):637–54.

Grathof, R. (ed.) (1978) *The Theory of Social Action, the Correspondence of Alfred Schutz and Talcott Parsons*, Bloomington and London: Indiana University Press.

Held, D. and J.B. Thompson (eds) (1990) *Social Theory of Modern Societies, Anthony Giddens and His Critics*, Cambridge: Cambridge University Press.

Holton, R.J. and B.S. Turner (1986) *Talcott Parsons on Economy and Society*, London: Routledge & Kegan Paul.

Holton, R.J. and B.S. Turner (1989) *Max Weber on Economy and Society*, London: Routledge.

Kronman, A.T. (1983) *Max Weber*, London: Edward Arnold.

Lepenies, W. (1985) *Die Drei Kulturen, Soziologie zwischen Literatur und Wissenschaft*, Munchen: Carl Hanser.

Levidow, L. and K. Robins (eds) (1989) *Cyborg Worlds, the Military Information Society*, London: Free Association Books.

Lidz, C.W. and V. Lidz (1976) 'Piaget's psychology of intelligence and the theory of action', pp. 195–239 in J.J. Loubser *et al.* (eds) *Explorations in General Theory in Social Science*, New York: Free Press.

Luckmann, T. (1957) *The Invisible Religion, the Problem of Religion in*

Modern Society, New York: Macmillan.

Maffesoli, M. (1989) *Au creux des apparences*, Paris; Plon.

Parsons, T. (1937) *The Structure of Social Action*, New York: McGraw-Hill.

Parsons, T. (1951) *The Social System*, London: Routledge & Kegan Paul.

Parsons, T. (1966) *Societies, Evolutionary and Comparative Perspectives*. Englewood Cliffs: New Jersey: Prentice-Hall.

Parsons, T. (1970) 'Some problems in general theory in sociology', in J. McKinney and E.A. Tiryakian (eds) *Theoretical Sociology*, New York: Appleton-Century-Crofts.

Parsons, T. (1978) *Action Theory and the Human Condition*, New York: Free Press.

Parsons, T. (1990) 'A tentative outline of American values', *Theory, Culture & Society* vol. 6(4):577–612.

Parsons, T. and E.A. Shils (eds) (1951) *Toward a General Theory of Action*, Cambridge, Mass: Harvard University Press.

Robertson, R. and B.S. Turner (1989) 'Talcott Parsons and modern social theory – an appreciation', *Theory, Culture & Society* vol. 6(4):539–58.

Runciman, W.G. (ed.) (1978) *Weber, Selections in Translation*, Cambridge: Cambridge University Press.

Sayer, D. (1989) *Readings from Marx*, London: Routledge.

Schutz, A. (1932) *Der sinnhafte Aufbau der sozialen Welt*, Vienna: Julius Springer. English translation (1972) by George Walsh and Frederick Lehnert, *The Phenomenology of the Social World*, London: HEB.

Schutz, A. (1970) *On Phenomenology and Social Relations*, Chicago: Chicago University Press.

Smith, J. and B.S. Turner (1986) 'Constructing social theory and constituting society', *Theory, Culture & Society* vol. 3:125–33.

Stauth, G. and B.S. Turner (1988) *Nietzsche's Dance, Resentment, Reciprocity and Resistance in Social Life*, Oxford: Basil Blackwell.

Thompson, J.B. (1990) 'The theory of structuration', pp. 56–76 in D. Held and J.B. Thompson (eds) *Social Theory of Modern Societies*, Cambridge: Cambridge University Press.

Turner, B.S. (1981) *For Weber, Essays in the Sociology of Fate*, London: Routledge & Kegan Paul.

Turner, B.S. (1983) *Religion and Social Theory*, London: Heinemann.

Turner, B.S. (1984) *The Body and Society, Explorations in Social Theory*, Oxford: Basil Blackwell.

Turner, B.S. (1987) *Medical Power and Social Knowledge*, London: Sage.

Chapter 3

Reflections on the epistemology of the hand

June 1933

> *Karl Jaspers* 'How do you think a man as coarse as Hitler can govern Germany?'

> *Martin Heidegger* 'Culture is of no importance. Look at his marvellous hands!'
>
> K. Jaspers (1977) *Philosophische Autobiographie*, Munich:101

INTRODUCTION: LAUGHING AND BARKING

Philosophical systems and social theories are often more profoundly tested at their periphery rather than at the core. Theories necessarily articulate a number of core or key issues, leaving marginal theoretical regions underdeveloped or underexplored. It is, therefore, in the soft under-belly of a theory that most of the critical damage may be done. In sociology, these marginal but sensitive regions have been called 'residual categories' (Parsons 1949). In the history of the theory of the sign, a similar problem has been identified by Umberto Eco and his colleagues:

> It is at its periphery that it gets put to the test. The Aristotelian theory of substance appears persuasive as long as one does not ask what the difference is between 'being a man' and 'being a boat'.
>
> (Eco *et al.* 1989:3)

The specific issue addressed by Eco is the problem of barking dogs within medieval theories of signs and symbols. The questions (Is a barking dog speaking? Is a laughing man talking?) exposed many confusions in the medieval conceptualization of signs, symbols, gestures and symptoms. Writers like Laurent Joubert in the six-teenth century attempted to preserve a sharp division between animal and human capacity for communication by claiming that laughter was exclusively human: 'For besides man, no animal laughs, unless perhaps it is a bastard laughter, simulated and counterfeit, such as those we call canine and sardonic' (Joubert 1980:94). But it proved difficult to settle on definitions of laughter and barking without making the whole theory of signs arbitrary. The small matter of a laugh threatened the whole system.

In this study, I have claimed that the unresolved status of the human body in relation to meaningful action is the residual category of action theory as a whole. The core of action theory has been the idea of voluntaristic selection between alternative courses of action. Because the analytical emphasis is put on rationality, normative evaluation and cognition, the embodied nature of the social actor is placed on the margin of the theory. In this chapter, I am trying to make this observation more specific by asking a naive and apparently trivial question: what is the role of the hand in human thought, gesture and interaction? Is shaking hands, waving hands, holding hands, binding hands, mutilating hands or cutting off hands of no sociological significance? Would it make no cultural difference if human beings had no thumb?

In autobiographical terms, this apparently trivial enquiry first occurred to me on reading Martin Heidegger, that most serious of twentieth-century philosophers, who has composed what one might regard as a eulogy on the meaning of the hand in volume fifty-four of the collected works, which deals with Parmenides (Heidegger 1982). I was drawn to this passage as a consequence of the debate over Heidegger's relationship to national socialism, specifically through Derrida's reflections on the hand in Heidegger's philosophy (Derrida 1989). Having awoken to the hand, the number of philosophers, anthropologists, sociologists and theologians who have written on the hand appeared to be suddenly extensive: Gaston Bachelard, Marc Bloch, Jean Brun, Aaron Cicourel, Robert Hertz, Marcel Mauss, G.H. Mead, Rodney Needham, Ovid, Quintillian, Erwin Straus and David Sudnow.

It now appeared that this peripheral question about a peripheral limb connected with many core issues in sociology: what is the relationship between animals and humans, between behaviour and action, and between nature and the social? This diversion into the hand also exposed an issue which lies behind most chapters in this collection, namely what, if any, is the connection between the physiology of the hand in relation to gesture and communication, and between an anthropological enquiry into the hand as a system of classification? Those writers who follow Robert Hertz (1960) are mainly concerned with the symbolic function of right-handedness in classificatory systems, and those sociologists who adopt a Meadian perspective *may* be more interested in the behavioural importance of the hand in relation, for example, to speech and thought. In short, we arrive at a contrast between a philosophical anthropology of the hand in the historical formation of human institutions and an analysis of the discursive importance of 'the hand' in classificatory systems.

ANIMALITY

The difference between humans and animals, and, more broadly, society and nature, has for many centuries exercised the imagination of philosophers and scientists. This quest for a contrast was not entirely innocent. Nature has been the mirror of society; either the noble savage has illustrated the degradation of 'civilized society', or nature appeared to be red in tooth and claw. Thus, Nature has often been employed discursively to justify the apparently innate character of 'Man' as competitive, aggressive and hierarchical (Haraway 1991; Hirst and Woolley 1982). Although there has been a general consensus that animals differ from humans in some crucial and unbridgeable respect, there has been little agreement about the precise character of that difference. While primate studies in physical anthropology and zoology have suggested that one crucial difference between human and animal spheres is the menstrual cycle and, as a consequence, permanent female receptivity, the social sciences have generally sought the difference between humans and animals in culture.

In cultural anthropology, Marxism and sociology, the main issue in making a distinction between animals and humans has involved an understanding of symbolic activity and tool use in the development of human civilization, that is between culture and technology. Human beings are believed to communicate uniquely through symbols and language, which are more than merely signs because language is a reflexively discursive form of communication. In a famous study of culture, Leslie A. White wrote that, although a dog can be taught to respond to 'Roll over!', a person 'can and does play an active role in determining what value the vocal stimulus is to have, and the dog cannot . . . the lower animals may receive new values, may acquire new meanings, but they cannot create and bestow them' (White 1949:29). What stands between 'Man' and animal is the whole world of culture, which civilized humanity in terms of directing sexual and aggressive behaviour towards altruistic and social activity (Elias 1978). Anthropology as the science of Man was in fact the science of culture (Malinowski 1960). In a similar fashion, Alfred Kroeber (1952) rejected evolutionism and argued that anthropology was the study of culture or the superorganic.

Culture, which is primarily constituted by the symbolic, has been taken as the essential dividing line between the biological and the social. The role of culture has been seen in terms of a system of restraints on behaviour, especially sexual behaviour. This notion of civilization as a regulation of egoistical sexuality gave rise to much scientific speculation about the crucial role of the incest taboo in prohibiting intra-familial sex, and thereby laying the foundation of family life (Parsons 1954). The civilizing process involved the development of codes of conduct, which normatively regulated aggressive behaviour. A long tradition of social theory from Nietzsche to Freud and Marcuse treated human society as a precarious balance between aggressive-sexual conduct (Dionysus) and social-rational arrangements (Apollo). Some image of homo-duplex has in fact dominated classical sociology (Stauth and Turner 1988).

Another version of these arguments puts a special emphasis on the adoption of tools as a crucial turning point in the history of humanity. In the case of Marxism, this idea should not be read as a form of technological determinism; it was an important feature of Marx's view of human beings as practical agents. The hand stands in a crucial relationship to this development. Hegel in the *Encyclopadie* had called the hand 'the tool of tools' by which men annex nature to their own bodily organs. The first tools were copies of human organs by which men extended their control over nature:

The use and fabrication of instruments of labour, although existing in the germ among certain species of animals, is specifically characteristic of the human labour-process, and Franklin therefore defines man as a tool-making animal.

(Marx 1974, vol. 1:175)

This vision of the dialectical relationship between Man as a sensual-practical being, nature and history was the basis of Marx's ontology. In *The German Ideology*, this view of human nature as practical activity received its classical expression:

> Men can be distinguished from animals by consciousness, by religion or anything else you like. They themselves begin to distinguish themselves from animals as soon as they begin to *produce* their means of subsistence, a step which is conditioned by their physical organisation.
>
> (Marx and Engels 1974:42)

In order to survive, human beings have to transform their natural environment through collective labour, but in this very process they begin to change their own nature. The development of human society thus occurs through a complex dialectic between 'the naturalization of man' and 'the humanization of nature'. Behind Marx's dialectical ontology of Man was an epistemological attack on idealism, which only grasped the subjectivity and consciousness of human beings and conventional materialism which regarded human beings as merely passive effects of organic processes. This position constituted the core of his attack on the legacy of Feuerbachian materialism, which failed to transcend the old dichotomy in German philosophy between passive idealism and active materialism. In Marx's ontology of praxis, human beings are sensuous, practical, embodied agents who, in the quest for survival, transform themselves in the course of transforming nature.

It is useful to contrast this image of Man in Marx and the sociology of Max Weber. Until recently it was unfashionable to attempt to develop a social ontology in the work of Weber (Turner 1981). There is, however, one possible link between Marx and Weber, which is that, in the theory of asceticism in the Protestant ethic, Weber outlined a social characterology. In the process of imposing an ethic of world mastery, the ascetic calling in the world also produced a special type of character (Hennis 1988). The world-mastery ethic sought to regulate the body, the mind and the environment through practices of restraint and discipline. In regulating the human body, the great quest of cultivated men (in Germany the *Bildungsbürgertum*) was to produce a particular type of person, namely a person who could subordinate the animal passions. This ethic was the principal target of Nietzsche's cultural critique. The quest for an ethic of ascetic responsibility produced the regulation and subordination of 'the world'. Weber's philosophical anthropology was buried in his account of the various soteriological pathways which had developed in the world religions, and thus Weber's view of the social dilemmas of human existence are to be found in his general account of theodicy. It is for that reason that I have put such an emphasis on the idea of theodicy in order to provide a theoretical space for the body in Weber's sociology of action – an idea which was first elaborated in my *Religion and Social Theory* (Turner 1983).

Many of these theories, which sought a difference between humans and animals in culture, symbol, or technology, have now become unpopular. They appear to be part of the useless debris of unworkable Victorian evolutionism. For example, Marx's view of human nature is criticized because the emphasis on labour is thought to be a paradoxical reflection of capitalist values of production. Marxism as a science is itself a form of fetishism because it reproduces the productivist values of industrial capitalism. In Baudrillard's work, such as *The Mirror of Production* (1975) and *Le système des objets* (1968), there is a sustained attempt to shift Marxist theory towards a better appreciation of the consumption of the sign as the basis for a radically different model of society. In a similar manner, Weber's attempt to give a privileged position to asceticism and rationality has been challenged as a patriarchal vision of human values (Bologh 1990). In more general terms, these conventional accounts of the human–animal divide have been made problematic by the recognition that some apes use tools, and that the boundary between animal sign language and human language is ambiguous. Other developments in genetic engineering, microbiology and the emergence of the cyborg have challenged the nature–society division.

Although there is now more disagreement than ever within the academic community over these previously taken-for-granted facts about the singularity of the human species, we can safely assume that within sociology there is a broad consensus around the ideas that (1) human nature is infinitely, or at least highly, malleable and plastic; (2) human nature is socially constructed, and as a consequence (3) there is little that one could say in *general* about human nature. Perhaps there is no need to say anything in general about human nature. The body, and especially the female body, is seen to be a 'fabrication' (Gaines and Herzog 1990) and the languages or discourses which describe the body (that is, medical, leisure, consumerist, artistic or legal discourses) are local, historical and specific perspectives. Thus, in the philosophy of science, there is now considerable support for the notion that nature is neither discovered nor uncovered; it is in fact constructed. The virtues and problems of social constructionism as an epistemology in medical sociology has been much disputed (Bury 1986; King 1987; Nicolson and McLaughlin 1987).

Although constructionism has much to recommend it, and although I do not dispute the idea that medical labels are as much interpretative devices as sociological labels, constructionism leaves open the problem, to which there is no neat answer, as to whether some things are more socially constructed than others. That is, one might adopt the position that, since everything is socially constructed (including the facts of natural science), the argument for constructionism is not very informative. It is in fact merely the starting point of the sociology of knowledge and not its conclusion. Most sociologists are likely to be weak constructionists. The argument about constructionism becomes more interesting in speaking of differences between disease categories in terms of *how* they may be differently constituted. Thus accepting constructionism might not necessarily entail anti-foundationalism, because some things are more constructed than others. We can distinguish between constructionism and anti-foundationalism, although in practice these positions are likely to be held simultaneously. For the sake of clarification, we can say that constructionism is a position within the sociology of knowledge, which claims that our knowledge of reality is the consequence of social processes. There are no discursively autonomous and neutral 'knowledges' of the world: the most 'concrete' facts about reality are social constructs. Anti-foundationalism is a form of social ontology which says that there are no 'things' or conditions which are not the product of social processes. There are no essential foundations outside ongoing social processes: the most concrete things are social products. Although it is possible in principle to distinguish these two positions, in general they tend to converge. Radical anti-foundationalists are likely to be radical constructionists.

It is also worth asking under what conditions sociologists themselves adopt or recommend constructionism. Typically, medical sociologists who adopt radical constructionism address themselves to politically sensitive or socially problematic disease entities. AIDS, trans-sexualism, repetitive strain injury, eating disorders, miner's lung, mental illness and psychosomatic illness generally are areas of constructionist research rather than goitre, gout or gonorrhoeal arthritis. Some 'disease categories' or 'sickness labels' such as anorexia nervosa, which are more socially constructed and overdetermined (Turner 1990) than other entities such as anaemia, are more socially contested than others. In short, social constructionism as an epistemological strategy tends to be invoked when there already exists a political struggle around the 'existence' of a disease; 'pre-menstrual tension' is a classic illustration.

In developing this argument, one might thus attempt to distinguish three issues. First, it may be valuable to make a distinction between 'conditions' and their descriptions. A 'condition' refers to any set of 'troubles' of individuals or groups which have yet to become the special concern of a professional ('helping') group and which have not been elaborately articulated by a discourse. In passing we might note that contemporary debates about constructionism have a lot in common with earlier approaches to 'social problems' as social processes; it is analytically related to arguments which assert that deviance is an effect of amplification and not a 'social fact' (Young 1971). Second, it is consequently possible to say that some conditions are more constructed than others; there is a continuum between a trouble and a well-established, respectable disease category. Finally, we might consider, as an exercise in the sociology of knowledge, when sociologists adopt constructionism. Let us provide an illustration of some aspects of the above set of positions.

The height of human populations can be measured over time with some accuracy as an indicator of environmental conditions and their impact on health. It is known that the average height of British working-class 14-year olds has increased by 29 cm over the last 250 years (Floud *et al.* 1990:184). There are many ways in which this condition of height might be described in social, natural or moral discourses. However, the notion of 'tall children' might at present be less socially constructed or overdetermined than the idea of the 'hyperactive child', because 'hyperactivity', while overtly a neutral psychological label, in fact might be a code for 'clever/troublesome'. A hyperactive child 'needs' special arrangements; tall children just need larger clothing. The sociological suspicion must be that hyperactivity is a myth (Schrag and Divoky 1975). However, 'tallness' might become an issue ('a trouble') if a social right was established to say that small people should not be discriminated against, for example in army recruitment practices or as candidates for vice-chancellorships. Under these circumstances, one could easily imagine a radical sociological critique which would demonstrate (to everyone's amazement) that 'height' is socially constructed.

These distinctions provide a pragmatic solution to some of the problems raised by constructionism. However, there is in sociology an implicit anxiety about 'essentialism', that is against the Aristotelian doctrine that some objects, however described, have essences. Heavily influenced by cultural relativism, sociologists are unwilling to make strong cross-cultural claims about 'human nature'. Perhaps the only minimalist assumption which still remains is that human beings are agents with a developed capacity to make and use symbols in order to sustain meaningful social relations. It is this capacity for symbolic action which lies behind the conventional sociological distinction between behaviour and action. The core assumptions of sociology, especially within the German tradition, have been that action involves the imputation of meaning and that social life is only possible on the basis of constant repair work to establish and defend common meanings.

Any attempt therefore to take the idea of a sociology of the body seriously, or at least to take the topic of embodiment as a focus of enquiry seriously, must adopt some position on these traditional debates about the boundary between humans and animals, and also take up some epistemological position *vis-à-vis* foundationalism and anti-foundationalism, between arguments for and against constructionism. If we argue that the Cartesian division between mind and body is spurious, and if we reject the privileged position traditionally given to instrumental rationality in defining the meaning of social action, does this mean that the conventional division between man and animals collapses?

CLASSIFICATION OF THE HAND

Before coming directly to this question, let us note that a number of social philosophers have addressed the question: is there some part of the body which is unique to human beings? That is, one might embrace the idea that, while apes and humans are embodied mammals, there is some crucial aspect of human embodiment which, through evolutionary specialization, has established a clear division between animals and humans. I have already drawn attention to the contentious view that it is the special reproductive process in human communities which separates them from animals. However, as a general rule, sociologists have not gone down this theoretical road; they have been mainly concerned with the symbolic significance of parts of the body. For example, the face has often been regarded as the critical aspect part of the human body because it is thought to reveal the soul. In a neglected essay, Georg Simmel explored the 'aesthetic significance of the face' (Simmel 1959), suggesting that it was the 'inner unity' of the face which was the key to its aesthetic. Any disfigurement could destroy this subtle unity. The face is the symbol of spirituality and personality. In modern societies, this emphasis on the face has been increased by the clothing of the body. While drawing attention to the spirituality of the face, Simmel noted that the hand is 'closest to the face in organic character' (Simmel 1959:276). Other writers, especially anthropologists, have been interested in the symbolic significance of hair and body decoration (Brain 1979; Hallpike 1969). In general, orifices have been regarded as crucial features of 'natural symbols' (Douglas 1973), because they are crucial in the symbolism of social membership as a consequence of entrances and exits. Body processes and the production of various excreta and fluids have also been regarded as essential features of human classificatory systems. Contact with human sperm, menstrual blood or faeces often has paradoxical consequences of transmitting health-giving charisma or disease and death. Christ's blood has played a major role in the ritualism and theology of Christianity, but Muslim saints also transmitted charisma (or baraka) through bodily fluids. The royal touch was regarded as a miraculous (charismatic) cure for the disfiguring disease of scrofula. The king's clothes were often thought to be efficacious in the cure of maladies (Bloch 1973). The exposed breast of the Virgin Mary in medieval religious art was a general symbol of care and nourishment in the context of famine and malnutrition (Miles 1986). Victor Turner (1966) has attempted to provide a general theory of the relationship between bodily processes, colours and classification and claimed that red (blood), black (excreta) and white (milk) are fundamental colours because they are associated with basic human experiences of the body, namely reproduction (red), defecation (black) and suckling (white). These anthropological approaches are based on a common assumption that the human body in general and parts of the body in particular function as elements within a system of classification.

This organizing principle of classificatory symbolism was the theoretical context within which Robert Hertz produced his justifiably celebrated essay on 'La prééminence de la main droite' in 1909 which appeared in English translation in *Death and the Right Hand* (Hertz 1960). This essay has been enormously influential in the development of anthropological theories of binary opposition in classificatory systems, especially the opposition between the right and the left (Needham 1973). One important feature of this essay, which in many ways continued the work of Durkheim and Mauss (1963), was the argument that the physiological asymmetry of the human body between left and right has merely a contingent relationship to the principle of left–right classification. The physiological tendency towards a preeminence of the right hand is only a pretext for classification.

Hertz recognized the existence of an anatomical basis of right-handedness, namely the greater development of the left cerebral hemisphere which activates the muscles of the opposite side. However, if right-handedness had an exclusively anatomical cause, how might we explain the almost universal social institutionalization of a cultural preference for right-handedness? This question is parallel to the problem of incest: if the natural aversion to incest is so great, why would there be such a strong social taboo (Fox 1967)? Why is it the case, for example, that in some societies such as the Netherlands Indies the left hand is bound to prevent its use? For Hertz, therefore, the preeminence of the right hand is a social institution, or in Durkheim's terms a social fact: the use of the right hand is positively sanctioned by society and deviation from right-handedness is often punished.

Following Durkheim's sociology of religion, Hertz regarded the polarity between right and left as a fundamental expression of the religious dualism between sacred and profane. This religious dualism came to shape the way in which social groups regard the body. The right hand represents the sacred side; it stands for male values, life, virility, power. The left hand represents evil, death, the female. In heraldry, the field of a shield was divided into left and right. A typical left-side design was the 'bend sinister', which gave us eventually the notion of sinister. A bend sinister or sinister bendlet was occasionally a sign of illegitimacy (Fox-Davies 1909:513). In contemporary societies, these divisions still exist. Right-handedness is associated with worthiness, dexterity, rectitude and beauty. In general, the hand is the basis of many ideas which embrace value judgements: handy; handsome; handicap; handful; high-handed. Traditionally a bargain (a hand-sale) was always sealed by a handshake. Right-handedness has assumed almost universal moral superiority. Christ entered Heaven to sit on the right-hand of God. The left side is occupied by demons and evil forces. Thus, Hertz concluded that 'the obligatory differentiation between the sides of the body is a particular case and a consequence of the dualism which is inherent in primitive thought' (Hertz 1960:110).

HEIDEGGER'S HAND

The results of this minor enquiry so far have been that interest in the hand in sociology and anthropology has been primarily concentrated in the symbolic functions of the hands, especially the right hand. My intention now is to begin to compare these approaches with phenomenology, philosophical anthropology and finally with Meadian symbolic interactionism. The point of this review is to begin to establish the claims of 'epistemological pragmatism' which will allow me to accept the procedures and discoveries of anti-foundationalism discourse analysis, while also retaining a foundationalist commitment to, for example, the philosophical anthropology of Gehlen and Plessner, and more recently the work of Berger and Luckmann.

In *Parmenides* Heidegger makes a number of extraordinary claims about the significance of the hand. An aspect of this claim depends upon the centrality of the word 'hand' in Old German. As we will see, Heidegger (and one might add Derrida's reflection on Heidegger) is able to move from the hand, to thinking, to technology and to being because of the etymological richness of the 'hand'. For example, while English has a similar metaphorical formulation, the relationship between 'to grasp' (*greifen*), 'to comprehend' (*begreifen*) and 'term' or 'concept' (*Begriff*) is probably more obvious in German. Heidegger's reflection on the hand is an extraordinarily imaginative meditation on *Hand, Handeln, Handwerk, Zuhandenheit* (readiness-to-hand) and *Vorhandenheit* (presence-at-hand).

Heidegger wants to make the bold assertion that animals do not have hands, and that a hand could not evolve from a paw or claw: *Kein Tier hat eine Hand, und niemals entsteht aus einer Pfote oder einer Klaue oder einer Kralle eine Hand* (Heidegger 1982:118). But Man has a hand only in a very special sense. Man has no hands but the hand stands for or occupies the place of man's being; it demarcates the field of man's being: *Der Mensche 'hat' nicht Hande, sondern die Hand hat das Wesen des Menschen inne, weil das Wort als der Wesensbereich der Hand der Wesensgrund des Menschen ist* (Heidegger 1982:119). The point of this expression is to bring Heidegger into the argument that speech and the hand are co-existing inseparable aspects of what it is that defines human beings (Heidegger obviously sticks to 'Man' and 'men') as human. Hand, gesture and speech cannot be separated, but the special function of the hand in the evolution of human societies is to reveal, to make manifest and to uncover thought in handwriting, or manuscripture. Heidegger as a consequence criticized the typewriter. The typewriter tears writing from the essential field of the hand; it degrades the word; and typing veils or masks the very essence of the activity of writing. Thus, prior to mechanization, hand, gesture, speech and writing stood in a relationship of integrated unity.

These reflections on hand/animal, society/technology, thinking/writing are part of a wider enquiry into the nature of being which is the central theme of Heidegger's entire philosophy. The reflection on the hand is a short-hand for his comparison of animate and inanimate nature (Derrida 1988). Thus, in *An Introduction to Metaphysics*, Heidegger (1959) claims that the stone has no world (*weltlos*), and the animal is deprived or poor in world (*weltarm*), while only man is world-forming (*weltbildend*). Heidegger's discussion of being is as a consequence related to the whole movement in German philosophical anthropology, because Heidegger presupposes, without specifically addressing, the 'findings' of such an anthropology. One linking figure between Heidegger and the philosophical anthropologists was Max Scheler whose work, such as *Man's Place in Nature* (1981), was influential in Heidegger's intellectual development. There are, however, a number of problems with the legacy of Heidegger from the perspective of sociology. For example, how do we move from notions of *Dasein* to ideas of collective existence, how do we move from presuppositions about zoology to a significant interaction between zoology, philosophical anthropology and sociology? Of course, Heidegger indicated a move from the analysis of *Dasein* and sociology via a relationship between *Hand* and *handeln*, which have the multiple meanings of 'bargain', 'behaviour', 'action' and 'to trade'. Heidegger played upon the idea of the handshake as the bond by which gift-giving and taking form organizations

and eventually societies. But of course it was not Heidegger's
intention to write a sociology of the hand, and it is not clear that
those sociologists (such as Scheler) whom he admired developed
a full sociological position on knowledge and being. For many
commentators, Scheler developed a phenomenology not a sociology
of knowledge (Frisby 1983). In order to make the step from a
phenomenology of the hand to a sociology of the hand, we need
to go via Arnold Gehlen to G.H. Mead.

ON HUMAN FRAILTY

Philosophical anthropology produced a rich and diverse literature
which attempted to incorporate the discoveries of biology and
zoology into the social sciences in a manner which avoided simple
reductionism or materialism (Honneth and Joas 1988). They
generally retained a strong sense of the very special place of 'Man'
in the universe and they were not indifferent to the traditional
arguments of the *Geisteswissenschaften* that the study of human action
required special approaches, but writers like Plessner wanted to
avoid the undimensional approach of materialism, psychologism
or vitalism to develop a concept of human unity as '*Leib-Geist-Einheit*'
(Plessner 1946). Their approach had its roots in Feuerbach's
sensualism, Marx's concept of praxis and human nature, in
Lebensphilosophie, in the existentialism of Søren Kierkegaard and in
phenomenology (Schnadelbach 1983). Arnold Gehlen, Helmuth
Plessner, Frederick J.J. Buytendijk, Paul Alsberg, Max Scheler
and Jakob J. von Uexkull made important contributions. In
contemporary sociology, this influence of philosophical anthropology
is most evident in the work of Peter L. Berger.

With considerable simplification, one might say that the core
of Gehlen's philosophical anthropology was a picture of human
beings – Gehlen never diverted from 'Man' – as frail and fragile.
A key idea in Gehlen's philosophy is expressed by *Entlastung*
(facilitation) and *belasten* (burdened). The verb has a similar range
of meaning in English. For example, a person may be burdened
by or weighed down with cares (*von Sorgen belastet*). *Entlasten* is to
relieve the strain or load on something. The noun *Entlastung* is relief,
which is translated as 'facilitation' in Gehlen's *Man in the Age of
Technology* (1980:3). Because men are born ill-equipped to deal with
the hostile environment of earth, they are burdened by the difficult
and constant necessity of providing for their own survival. Human

beings do not live their lives; they have to lead them out of necessity. Man is forced to lead, not live, 'not for reasons of enjoyment, not for the luxury of contemplation, but out of sheer desperation' (Gehlen 1988:10). Gehlen argued on the basis of comparative embryology that the period of human gestation was too short; human beings develop after birth by a long period of nourishment, socialization and care. Man is 'unfinished' (Gehlen 1988:4). The institutions, practices, cultural arrangements and social structures which help humans to overcome these limitations (by providing relief) are illustrations of *Entlastung*.

Gehlen defined the human being as an 'acting being' (*handelndes Wesen*); man is compelled to action out of the very incompleteness of being. Here again we should not miss the relationship between *handelndes* and *hand*. Humans are 'naturally' oriented towards the future rather than the present. An active being who is future-directed needs to be handy; humans need things to be available (*vorhanden*) for planning. Humans must anticipate future possibilities through acts of imagination, and they must discipline themselves now in order to be ready to grasp future opportunities.

This unfinished character – organic primitiveness and the absence of natural means of coping – of humans has a number of components. Humans are described in terms of their 'world-openness', that is they do not have specific instincts related to the satisfaction of needs in a given environment. In fact, human beings are poor in instinctual equipment; they experience 'instinctual deprivation' (*Instinktarmut*). Because humans cannot depend on a limited and specific range of instincts to provide a selective link to the outer world, humans are burdened by overwhelming stimulation. This vulnerable animal requires both the relief of strong institutionalization to survive, and cultural selectivity to manage the overload of impulses which are biologically unstructured. As a result 'it is possible to define man as a being of discipline' (Gehlen 1988:52).

The hand plays an important part in these arguments. Because humans are frail and unfinished, they have compensated for this weakness by organ substitution. The stone weapon substitutes for the hand as a relatively weak fighting organ. The origin of technology or hand-work is a set of processes related to human incompleteness: replacement, strengthening and facilitation techniques. The replacement of the organic by the inorganic is a fundamental principle of human cultural development. But the hand is also crucial to the exploration of the environment and crucial to

humans as beings of action who are forced to project themselves
into the future. The hand thus comes into play with imagination
and speech. The exact coordination of hand and eye is the foundation of human action. There is a circulary movement between
hands, eyes and language, which arises out of endless practice and
discipline. Gehlen's theory of speech, however, is not intellectualist,
because speech is not so much the articulation of thoughts; speech
involves acting without acting.

It is here that there is an obvious connection with the work of
Mead. As Miller (1984:60) has argued, few commentators on Mead
have noticed the centrality of the hand to Mead's account of the
origins of reflective thinking in human beings. In lower animals
there are two phases to action: the stimulus phase sets free an
impulse which is followed by a consummatory phase. A tiger smells
its prey and proceeds to attack and devour it. In humans there is
an intervening phase – the manipulatory phase – when the object
is manipulated, explored and touched by the hand. Mead argued
in his unpublished lectures on social psychology in spring 1927 that
mind came into function in the intermediary phase between
stimulation and consumption. It is in this intermediary/manipulative
phase that mind can imagine and consider alternative courses of
action. Reflective thinking is made possible by this space, and this
space is a consequence of the dexterity of the hand, which helps
to liberate or break up fixed instincts by offering humans a world
full of possibilities. In short, the hand is, as Gehlen would say, an
important aspect of human world-openness. The origin of language
is thus to be located in the work of the hands, once liberated from
supporting the body, in grasping, manipulating, breaking up and
reassembling the objects of our immediate environment:

> mind is an emergent, even as is the hand, and there is a
> functional relationship between the two. We see what we
> handle and we handle what we see, and we know the world in
> 'handfuls'.

> (Miller 1984:62)

Because our hands are instruments, they play a major part in
speculation about means and ends. The hand is an esssential feature
of instrumental rationality (Mead 1934:245–9). This manipulative
stage is also important for the human ability for empathy, for
imagination in role play, and hence for the origins and development
of social behaviour.

THE BODY AND THE SOCIAL CONSTRUCTION
OF REALITY

Gehlen's anlaysis has been both influential and contentious. He diagnosed the problem of the modern world as one of deinstitutionalization, and produced a conservative theory of discipline and institution as essential for human stability. These political aspects of Gehlen's work are not relevant here. Instead let us focus on an analytical problem in Gehlen, namely how to produce a theory of society from an anthropology of human frailty. The main problem with Gehlen's philosophical anthropology is its individualism, because he connects it 'so resolutely and exclusively to an individualist model of action that he is unable to become aware of the significance of intersubjective modes of action for socialisation as well as for the history of the human species' (Honneth and Joas 1988:58). One solution to this problem would be the adoption of a Meadian perspective on intersubjectivity and the concepts of role-taking, the generalized other and the self. Of course, Gehlen himself had indicated the value of Meadian social psychology as an addition to his own approach. There are good reasons why Mead and Gehlen might be integrated to form a more general theory. As Honneth and Joas (1988:61) correctly indicate, in interpreting Mead as a founder of symbolic interactionism, sociologists have often ignored the subtitle ('from the standpoint of a social behaviorist') of Mead's famous text on *Mind, Self and Society*. In fact, Mead's theory developed around a fundamental interest in the relationship between organism and environment. Mead defined the 'mechanism of human society' as that 'of bodily selves who assist or hinder each other in their co-operative acts by the manipulation of physical things' (Mead 1934:169). The problem then is how to achieve intersubjectivity on the basis of Gehlen–Mead anthropology-behaviouralist presuppositions. The answer is through an intersubjective theory of communication, which does not deny the achievements of philosophical anthropology. Honneth and Joas appear to be correct in pointing to Elias's civilization-process thesis and to the historical anthropology of the family as possible lines of development in the legacy of Gehlen and Mead. It is both interesting and curious that Honneth and Joas do not take account of the development of the tradition of philosophical anthropology through the work of Berger and Luckmann into contemporary debates about the social construction of reality.

In the late 1970s, sociology in Britain and North America was profoundly influenced by the publication of *The Social Construction of Reality, a Treatise in the Sociology of Knowledge* (Berger and Luckman 1966). This development of the sociology of knowledge had a profound impact on various fields of sociology, but especially in the sociology of religion (Berger 1967; Luckmann 1967), where Berger argued that 'the sacred canopy' of religion was erected as a defence against the terror of chaos and anomie. However, the basic arguments of the sociology of knowledge were widely applied to marriage (Berger and Kellner 1964), personal identity (Berger and Luckmann 1964) and psychoanalysis (Berger 1965). In the 1970s, Berger's work, which offered a critique of deterministic theories in the sociology of knowledge, was regarded as a radical attack on existing social structures. The constructionist aspect of Berger's sociology promised an alternative theoretical framework, restoring human agency and implying the political possibility of individuals acting against powerful social structures such as bureaucracies (Abercrombie 1986: 11).

Berger was critical of classical sociology on two counts. First, it concentrated too heavily on articulate, literate and intellectual traditions rather than on 'knowledge' as such. Writers like Mannheim were concerned to understand belief *systems* such as conservatism and liberalism, but they failed to analyse everyday, taken-for-granted meanings. Berger thus attempted to understand everything that passes for knowledge in everyday life. Secondly, he criticized the deterministic legacy in classical sociology and developed his own view of the dialectical relationship between externalization, objectivation and internalization. Thus, social human beings create their own world (because they are world-open), they bestow a sense of objectivity and permanence on these human products (because they cannot live with uncertainty) and they reappropriate this external reality as part of their own subjectivity (because they need to live in a world which is meaningful and plausible). It was in terms of this dialectical picture that Berger wanted to reconcile the relationship between human agency and social structures.

Critics of Berger have often thought that this solution to the problem of structuration does not work, and that in fact Berger is forced back into a view of human agency as determined by unmovable social structures and cultural necessities. For example, the 'anthropological constancies' in Berger's theory 'represent *limitations on autonomous human activity*. Together they give an oppressive feeling of weight' (Abercrombie 1986:29). Abercrombie's criticism is clearly appropriate but he failed to recognize that Berger's 'anthropological constancies' are taken directly from Gehlen's philosophical anthropology (Berger and Kellner 1965). Berger's social actor does indeed feel the weight of their requirements because human beings are burdened creatures who require the relief (*Entlastung*) of culture. Berger thus embraced Gehlen's notion of Man as by nature a creature of action because they are compelled to forward-looking actions in order to survive. The paradox of this position is that human agency is borne out of biological necessity; hence Berger's social agent is simultaneously free and determined, which explains how Berger can embrace the dialectical paradox of externalization, objectivation and internalization: human beings create society, and are created by society. Berger combined this social ontology with the phenomenology of Alfred Schutz (1971) to argue that the sociology of knowledge should be concerned with the 'facticity' of the everyday world, because it is at this everyday level that the sacred canopy operates to block any awareness that reality is socially constructed. Again the paradox here is the following: all reality is socially constructed, as a consequence of Man's incompleteness, but human beings require stable meanings and cannot live in permanent awareness of the socially constructed and precarious nature of everyday reality, and they are forced to clothe these uncertainties with permanent significance. The precarious nature of the continuously-socially-constructed-world is disguised by the sacred canopy of shared realties. This reality-formation is proved by religion.

Although I do not support Berger's approach in its entirety, he provides a solution to epistemological problems that is very congenial to the approach I have suggested in general towards the sociology of the body. Berger's position is unusual but defensible, namely a foundationalist ontology combined with a constructionist view of knowledge. Thus, Berger takes on Gehlen and Plessner's foundationalist view of human embodiment within a broadly evolutionist framework, which he combines with a view that our knowledge of reality is socially constructed. This view also appears to be shared by Norbert Elias in his *The Symbol Theory* (1991), where Elias argues that symbols are also tangible patterns of sound in social communication and as such are made possible by the evolutionary development of the vocal apparatus. Similarly, Berger's argument is that, although culture is certainly socially constructed, its construction is the consequence of the very peculiar and unique biological foundations of the human animal as 'a not yet determined creature'.

My modifications of Berger are along two lines. First, when Berger refers to 'reality', he appears to mean 'social reality', whereas a radical constructionist view of knowledge would say that knowledge of all reality (natural and social) is socially produced. This view would be supported by, for example, the work of Ludwik Fleck (1979) who demonstrated that scientific facts are the products of what he called 'thought communities'. This point is merely to note that Berger's radical sociology of knowledge should also be applied to scientific discourses. Secondly, Berger apparently believes that all knowledge is in some sense equally constructed; there is no variation. This position is, in my view, a mistake. By arguing for the possibility that some conditions or circumstances might be more socially constructed than others, we also leave open the possibility of ideological critique, that is we leave open the possibility of political action to deconstruct existing social constructions.

We can now see that is it possible to hold a foundationalist view of the significance of the human hand in the evolution of culture and society, and a notion that the 'hand' is a discursive construct within a classificatory paradigm which is fundamental to human society, namely the idea of the superiority of the right hand. The basic notions of goodness and evil are bound up with the fact that left-handedness is a sinister accomplishment. However, the fundamental physiological feature of the hand is dexterity, which is closely associated with the flexibility of the thumb. It is on this physiological basis that human culture has developed endless cultural complexity. Playing the piano might be one rather obvious illustration of the complex interaction between potentialities, training, discipline and culture (Sudnow 1978). The body provides the foundational potentialities upon which endless cultural practices can be erected.

REFERENCES

Abercrombie, N. (1986) 'Knowledge, order and human autonomy', pp. 11–30 in J.D. Hunter and S.C. Ainlay (eds) *Making Sense of Modern Times, Peter L. Berger and the Vision of Interpretive Sociology*, London: Routledge & Kegan Paul.

Bachelard G. (1988) 'Hand vs. matter', pp. 51–3 in *The Right to Dream*, Dallas: Dallas Institute Publications.

Baudrillard, J. (1968) *Le système des objets*, Paris: Gallimard.

Baudrillard, J. (1975) *The Mirror of Production*, St Louis: Telos Press.

Berger, P.L. (1965) 'Towards a sociological understanding of psychoanalysis', *Social Research* vol. 32.

Berger, P.L. (1967) *The Sacred Canopy*, Garden City, New York: Doubleday.

Berger, P.L. and H. Kellner (1964) 'Marriage and the construction of reality', *Diogenes* vol. 46: 1–24.

Berger, P.L. and H. Kellner (1965) 'Arnold Gehlen and the theory of institutions', *Social Research* vol. 32: 110–15.

Berger, P.L. and T. Luckmann (1964) 'Social mobility and personal identity', *European Journal of Sociology* vol. 15: 331–44.

Berger, P.L. and T. Luckmann (1966) *The Social Construction of Reality, a Treatise in the Sociology of Knowledge*, Garden City, New York: Doubleday.

Bloch, M. (1973) *The Royal Touch, Sacred Monarchy and Scrofula in England and France*, London: Routledge & Kegan Paul.

Bologh, R.W. (1990) *Love or Greatness: Max Weber and Masculine Thinking – a Feminist Inquiry*, London: Unwin Hyman.

Brain, R. (1979) *The Decorated Body*, London: Hutchinson.

Bury, M.R. (1986) 'Social constructionism and the development of medical sociology', *Sociology of Health & Illness* vol. 8(2): 136–69.

Derrida, J. (1988) 'Geschlecht 11: Heidegger's Hand', pp. 161–96 in John Sallis (ed.) *Deconstruction and Philosophy, the Texts of Jacques Derrida*, Chicago and London: Chicago University Press.

Derrida, J. (1989) *Of Spirit, Heidegger and the Question*, Chicago and London: Chicago University Press.

Douglas, M. (1973) *Natural Symbols*, New York: Vintage.

Durkheim, E. and M. Mauss (1963) *Primitive Classification*, London: Cohen & West.

Eco, U., R. Lambertini, C. Marmo and A. Tabarroni (1989) 'On animal language in the medieval classification of signs', pp. 3–41 in U. Eco and C. Marmo (eds) *On the Medieval Theory of Signs*, Amsterdam and Philadelphia: John Benjamins Publishing Co.

Elias, N. (1978) *The Civilising Process*, Oxford: Basil Blackwell.

Elias, N. (1991) *The Symbol Theory*, London: Sage.

Fleck, L. (1979) *Genesis and Development of a Scientific Fact*, Chicago: Chicago University Press.

Floud, R., K. Wachter and A. Gregory (1990) *Height, Health and History. Nutritional status in the United Kingdom 1750–1980*, Cambridge: Cambridge University Press.

Fox, R. (1967) *Kinship and Marriage*, Harmondsworth: Penguin Books.

Fox-Davies, A.C. (1909) *A Complete Guide to Heraldry*, London: T.C. & E.C. Jack.

Frisby, D. (1983) *The Alienated Mind, the Sociology of Knowledge in Germany 1918–1933*, London: Heinemann Educational Books.

Gaines, J. and C. Herzog (eds) (1990) *Fabrications, Costume and the Female Body*, London: Routledge.

Gehlen, A. (1980) *Man in the Age of Technology*, New York: Columbia University Press.

Gehlen, A. (1988) *Man, His Nature and Place in the World*, New York: Columbia University Press.

Hallpike, C.R. (1969) 'Social hair', *Man* vol. 4: 256–64.

Haraway, D.J. (1991) *Simians, Cyborgs and Women: the Reinvention of Nature*, London: Free Association Books.

Heidegger, M. (1959) *An Introduction to Metaphysics*, New Haven: Yale University Press.

Heidegger, M. (1982) *Gesamtausgabe, Band 54, Parmenides*, Frankfurt am Main: Vittorio Klostermann.

Hennis, W. (1988) *Max Weber, Essays in Reconstruction*, London: Allen & Unwin.

Hertz, R. (1960) *Death and the Right Hand*, London: Cohen & West.

Hirst, P. and P. Woolley (1982) *Social Relations and Human Attributes*, London: Tavistock.

Honneth, A. and H. Joas (1988) *Social Action and Human Nature*, Cambridge: Cambridge University Press.

Jaspers, K. (1977) *Philosophische Autobiographie*, Munich.

Joubert, L. (1980) *Treatise on Laughter*, Alabama: Alabama University Press.

King, D. (1987) 'Social constructionism and medical knowledge: the case of transsexualism', *Sociology of Health & Illness* vol. 9(4):351–77.

Kroeber, A.L. (1952) *The Nature of Culture*, Chicago: Chicago University Press.

Leder, D. (1990) *The Absent Body*, Chicago and London: Chicago University Press.

Luckmann, T. (1967) *The Invisible Religion, the Problem of Religion in Modern Society*, New York: Macmillan.

Malinowski, B. (1960) *A Scientific Theory of Culture and Other Essays*, New York: Oxford University Press.

Marx, K. (1974) *Capital*, London: Lawrence & Wishart, 3 vols.

Marx, K. and F. Engels (1974) *The German Ideology*, London: Lawrence & Wishart.

Mead, G.H. (1934) *Mind, Self and Society: From the Standpoint of a Social Behaviorist*, Chicago: Chicago University Press.

Miles, M.R. (1986) 'The Virgin's One Bare Breast: female nudity and religious meaning in Tuscan early Renaissance culture', pp. 193–208 in S.R. Suleiman (ed.) *The Female Body in Western Culture, Contemporary Perspectives*, Cambridge, Mass: Harvard University Press.

Miller, D.L. (1984) *George Herbert Mead, Self, Language and the World*, Chicago and London: Chicago University Press.

Needham, R. (ed.) (1973) *Right and Left, Essays on Dual Symbolic Classification*, Chicago and London: Chicago University Press.

Nicolson, M. and C. McLaughlin (1987) 'Social constructionism and medical sociology: a reply to M.R. Bury', *Sociology of Health & Illness* vol. 9(2):107–26.

Parsons, T. (1949) *The Structure of Social Action*, New York: Free Press.

Parsons, T. (1954) 'The incest taboo in relation to social structure and socialization of the child', *British Journal of Sociology* vol. 5(2): 101–7.

Plessner, H. (1946) 'Mensch und Tier', pp. 302–17 in Leroy E. Loemker (ed.) *Gottfried Wilhelm Leibniz*, Hamburg: Hansischer Gildenverlag.

Scheler, M. (1981) *Man's Place in Nature*, New York: FS & G.

Schnadelbach, H. (1983) *German Philosophy 1831–1933*, Cambridge: Cambridge University Press.

Schrag, P. and D. Divoky (1975) *The Myth of the Hyperactive Child and Other Means of Child Control*, New York: Pantheon Books.

Simmel, G. (1959) 'The aesthetic significance of the face', pp. 276–81 in Kurt H. Wolf (ed.) *Essays on Sociology, Philosophy and Aesthetics*, New York: Harper & Row.

Stauth, G. and B.S. Turner (1988) *Nietzsche's Dance: Resentment, Reciprocity and Resistance in Social Life*, Oxford: Basil Blackwell.

Sudnow, D. (1978) *Ways of the Hand, the Organization of Improvised Conduct*, London and Henley: Routledge & Kegan Paul.

Turner, B.S. (1981) *For Weber, Essays in the Sociology of Fate*, London: Routledge & Kegan Paul.

Turner, B.S. (1983) *Religion and Social Theory, a Materialist Perspective*, London: Heinemann Educational Books.

Turner, B.S. (1990) 'The talking disease: Hilda Bruch and Anorexia Nervosa', *Australian and New Zealand Journal of Sociology* vol. 26(2):157–69.

Turner, V.W. (1966) 'Colour classification in Ndembu ritual', in M. Banton (ed.) *Anthropological Approaches to the Study of Religion*, London: Tavistock.

White, L.A. (1949) *The Science of Culture, a Study of Man and Civilization*, New York: Grove Press.

Young, J. (1971) *The Drugtakers, the Social Meaning of Drug Use*, London: Paladin.

Part II

Medical sociology

Chapter 4

The interdisciplinary curriculum
From social medicine
to postmodernism

INTRODUCTION: DEFINITIONS

Although experiments in the reorganization of the medical curriculum have been a constant feature of higher education systems in the postwar period (Bloom 1988; Light 1988), universities currently face acute financial and organizational problems, which in turn have major implications for professional education and professional autonomy (Abbot 1988). The reorganization of higher education systems is significant not only for separate disciplines, but for the relationships between disciplines. Indeed, one major component of contemporary curriculum development is the plea for greater interdisciplinarity, which is typically combined with the demand on the part of governments for greater social relevance and problem orientation. Interdisciplinarity has emerged in a context where it is claimed that contemporary health (or more generally social) problems cannot be tackled on a monodisciplinary basis; the interdisciplinarity debate is tied therefore to the quest for effective 'problem-solving'. Although the notion of interdisciplinarity as a general objective of education reform is contentious (Kocka 1987; Piaget 1970), there is some agreement (however minimal) that the scientific study of health and illness is an area which is peculiarly suited to an interdisciplinary approach. For example, the British Open University course on health and illness is based on the assumption that 'Neither health nor disease are straightforward matters, and that they can only be fully understood by adopting an interdisciplinary stance' (Black *et al.* 1984: xi). Similarly, it can be argued that 'health and illness is an area which, theoretically, is ripe for fruitful interdisciplinary efforts' (Charmaz 1986: 279). While these claims have a *prima facie* validity, we need a more elaborated notion of interdisciplinarity in order to understand why it may be more theoretically fruitful than conventional monodisciplinary approaches. What is interdisciplinarity?

For heuristic purposes, let us argue that the social organization of the sciences can be conceptualized in terms of a hierarchy of growing complexity: disciplinarity, multidisciplinarity and interdisciplinarity. Although I shall subsequently challenge this view, we may define a discipline as a more or less coherent study of a topic or field from a more or less unitary perspective. This definition is deliberately minimalist, because the outcome of this paper is to suggest that what look like coherent disciplines turn out to be typically loose affiliations or federations of theories, perspectives, topics and methods which could be easily redistributed within the university system. Multidisciplinarity is simply a collection of such disciplines which are assembled for the study of a topic or range of topics. There is no necessary attempt to produce a coherent assembly or a theoretically systematic regrouping of existing disciplines. Whereas monodisciplinarity is an *ad hoc* assemblage, interdisciplinarity aims in principle at academic fusion. Interdisciplinarity, because it seeks a reorganization and integration of disciplines, involves a critique of disciplinary practices. Because interdisciplinarity challenges the organization of the conventional curriculum, it ultimately raises questions about the professional division of labour in health-care systems. This account of the nature of interdisciplinarity is not merely descriptive; it contains implicit normative views about academic change. Given the complexity of health issues, the approach of medical and social sciences *ought* to be interdisciplinary. However, in this chapter I attempt to contrast (what we may call) positive and negative forms of interdisciplinarity. The positive case (for example, social medicine) is based on some theoretical principles, which stipulate interdisciplinarity as a necessary basis of the curriculum for reasons which are broadly scientific. The negative example (which is referred to in this chapter as the McDonaldization of the curriculum) is the unintended consequence of changes in the organization of research and teaching, which are brought about for reasons which are broadly economic.

In order to go beyond a merely monodisciplinary approach to create a genuinely integrated interdisciplinary field, it is necessary for interdisciplinarity to adopt an epistemologically creative and critical stance towards existing disciplinarity. For example, one justification for an interdisciplinary reorganization of both medical and social science faculties would be that an adequate scientific approach to health and illness requires an understanding of the complex causality of illness and disease, and that a valid therapeutics must be grounded in a holistic view of the patient. Thus, the claim that an interdisciplinary approach is essential for the development of medical science will come to depend eventually on what we mean by 'complexity'. An interdisciplinary approach will have to develop a fairly sophisticated epistemology of disease which would entail at least the following. Interdisciplinarity requires reflexivity, that is an awareness of the historical and social setting of scientific concepts. A sociology of knowledge of health and disease entities is a necessary feature of such an epistemology (Turner 1987). This reflexivity would lead either to a relativistic view of disease entities or to some notion of social constructionism (Bury 1986). A constructionist epistemology throws doubt upon the idea of theory-neutral medical facts and more importantly casts doubt upon the idea of unambiguous medical progress. For example, Michel Foucault, following Gaston Bachelard's concept of the 'epistemological rupture' (Bachelard 1934), has made us familiar with the idea of major discontinuities in the development of scientific knowledge. The recent revival of interest in the work of Ludwik Fleck (Cohen and Schnelle 1986) has drawn sociological attention to the notion that the emergence of a scientific fact is the effect of various thought-styles (*Denkstil*) which in turn are supported and maintained by a thought collective (*Denkkollektiv*). The epistemology of an interdisciplinary approach is sceptical with respect to professional and other claims to truth. Interdisciplinarity, for example, tends to be sceptical as to the claims of scientific medicine and the medical model, and it is based on the notion of the essential multicausality of social, individual, biological and cultural phenomena. This critical epistemology implies a reorganization of the medical curriculum, the transformation of the relationship between medicine and the academy, and a different relationship between doctors and patients. Interdisciplinarity will inevitably involve conflictual professional relationships. These occupational conflicts are in part a function of the tension between the aspiration for interdisciplinarity and the 'solidarity of the medical profession' (Strong 1984: 346). In the final analysis, interdisciplinarity throws doubt not only on the professional claims of medical science, but on the character of the disciplinary division of the social sciences.

The final component of interdisciplinarity which we need to consider is the question of the problem-focused character of such research. This orientation to problem-solving is the most uncertain component of the interdisciplinary complex. By way of over-simplification, we can note that historically clinical medicine was focused on the discomfort (that is the disease) of the individual patient. Empirical medicine has taken a specifically hostile stance towards the theoretical justification of adequate clinical practices, and there has been characteristically a division between experimentally based medicine, medicine within the university and the clinical practice of the general practitioner. For example, Sydenham and Locke were critical of experimental pathological anatomy on the grounds that it was dominated by an abstract theoretical enquiry which was irrelevant to the day-to-day practice of medicine and the management of patients. The empiricist revolution of the seventeenth century against the theoretical orientation of Galenic medicine left a legacy in which there is a divorce between theoretical enquiry in the natural sciences and a clinically based medical practice (King 1982). The development of this positivistic problem-oriented medicine resulted in a specialization of medical disciplines around various parts of the human body, thereby excluding on professional and scientific grounds the claims of a holistic approach to medical practice, at least until the 1960s when there was a revival of so-called biopsychosocial medicine in the United States (Gordon 1984). We may assert at this stage that a radical interdisciplinary approach is related to the notion of holistic medicine, and both regard the idea of a medical problem as itself problematic. The creation of a comprehensive interdisciplinarity within the social sciences, and between medicine and social science, requires some reconciliation between an atheoretical clinical practice, the theoretical development of the fundamental sciences which underlie medical knowledge, and the applied social sciences. In the historical development of modern medicine, various attempts have been made, either explicitly or implicitly, to achieve some or all of the goals of interdisciplinarity within and between the medical and social sciences. In this overview of some aspects of the development of interdisciplinarity, the problems and prospects of change in the medical curriculum will be reviewed through four examples, starting with the most general, namely the idea of a social medicine.

SOCIAL MEDICINE

The concept of 'social medicine' has had a complex and changing history, referring at different times to very different practices (Porter and Porter 1988). However, despite these various definitions and meanings, the idea of social medicine provides us with an important historical precursor for the development of interdisciplinary social medical sciences. Social medicine had its origins in the eighteenth century and was associated with the development of greater state intervention under the doctrine of mercantilism. George Rosen (1979) in an important article on 'the evolution of social medicine' notes that the idea of public intervention in health matters arose in the context of the development of police science (*Polizeiwissenschaft*) within which health administration was seen to be an important part of the general policing of society, giving rise to the idea of a medical police (*Medizinalpolizei*). For example, between 1779 and 1817 Johan Peter Frank produced his six-volume analysis of the government interventions which would be necessary for the protection of individual health within a social context (*System einer vollständigen medizinischen Polizei*).

Frank advocated a public health policy which, in its surveillance of the population, was paternalistic and authoritarian. However, it did pioneer the development of a thorough and systematic approach to the health problems of social life, but it was in France in the nineteenth century that a more theoretically sophisticated advance in social medicine was finally established. In the urban crises following the industrialization of France in the late nineteenth century, and as a long-term consequence of the political disturbances of the French Revolution, the influence of St Simonian social reformism was fully experienced. It was Jules Guèrin (1801–86), the editor of the *Gazette Medicale de Paris*, who developed the term 'social medicine' as a consequence of the innovative public surveys of Nantes in 1835, which employed new statistical methods of survey analysis to understand the extent of public illness. Guèrin in *Médecine sociale au corps médicale de France* (1848) divided social medicine into social physiology, social pathology, social hygiene and social therapy. Guèrin gave particular emphasis to the social and political functions of the physician's role in a revolutionary context where he called for an organized medical intervention into all social spheres. It is not surprising therefore that Foucault saw in these medical surveys and medical programmes the true origins of modern sociology, and he dismissed Montesquieu and Comte as the founders of a scientific science of society (Foucault

1980: 151). It was this 'accumulation of men' which produced 'population' as the great 'object of surveillance, analysis, intervention, modification' (Foucault 1980:171; Turner 1985). My proposal is that this conception of social medicine implied an interdisciplinary approach to the political management of populations in terms of their health and general social requirements, but that it was also part of a systematic critique of the conventional role of the medical man in society.

In Germany these revolutionary French ideas were converted to a self-conscious critical social science of human medical problems, and in particular it was Rudolf Virchow who harnessed medical science to the political transformation of European societies (Ackerknecht 1953). In his famous report on the typhus epidemic in Upper Silesia in 1847–48, Virchow developed the argument that the causes of this epidemic were as much social, political and economic as biological and physical in character; the health of communities could ultimately only be firmly improved as a consequence of major political, social and environmental reform, including the democratization of the political system. In short, the health problems of society could only be resolved as a consequence of radical intervention based on an interdisciplinary approach, and he developed the radical slogan that 'medicine is a social science, and politics nothing but medicine on a grand scale' (Virchow 1848:2). Virchow understood that health could only be improved by socioeconomic as well as medical interventions, but equally he recognized that the impediments to such intervention were also political in their essence, and therefore he has to some extent anticipated the radical political economy of health represented in the work of writers like Vincente Navarro (Taylor and Rieger 1984). However, the defeat of radical politics in 1848 in both Germany and France brought about a temporary halt to the movement for a radical interdisciplinary health programme in social medicine, but many of Virchow's ideas were either implicitly or explicitly reproduced elsewhere in Europe. For example, in England, Edwin Chadwick's *Enquiry into the Sanitary Conditions of the Labouring Population in 1842* and Friedrich Engels's *Condition of the Working Classes in England in 1845* came to rather similar conclusions as to the importance of social medicine.

From this brief sketch of the early history of social medicine, we may derive three characteristics of this movement. First, there was, particularly in the work of Virchow, the recognition that illness has to be understood in multicausal terms. Second, to understand and therefore to change the nature of the health status of a

population, it is essential to undertake social and political intervention and reform. Third, and as a consequence to the first two, social
medicine emerged as a radical political movement which was critical,
not only of the intervention of traditional medicine but of the entire
society. It is not an accident that social medicine emerged in
response to and as a consequence of the French Revolution and
later the revolutionary conflicts of 1848. René Sand in his *Vers la
Médecine* sought the origins of social medicine in broader social
changes (such as the emergence of social insurance, the institutional
development of the general hospital and the expansion of social
sciences), but it is interesting that social medicine has often been
referred to, or been confused with, the idea of socialized medicine.
Henry Sigerist (1937) in particular treated social medicine as
socialized medicine and developed a clear political conception of
the operation of a free public health system based on general taxation which he thought would emancipate medicine from the
economic constraints of a competitive capitalist economy. In a
similar fashion, George Rosen in his study on public health also
suggested that the medical man was the natural ally of the poor
and that medicine had, as it were, a natural function in social
amelioration.

Of course, there is no *necessary* connection between social, preventive or socialist medicine, and indeed in the English context, when
interventionist medicine was combined with eugenics, then a reaction-
ary doctrine emerged in which the state was involved in the biological
planning of the community through the development of a mechanism
for selective breeding. The eugenic ideology provided a new basis
for the state to institute a total government of the body (Turner 1982).
However, both left-wing and right-wing versions of social medicine
had one thing in common, a particular view of the state:

> Social medicine depended on scientifically informed,
> technocratically determined actions by the state. This technocratic
> vision differentiates the ideas of social medicine from theories
> of socialist medicine in which the vision of the state is political,
> not technical. The latter looks for the causes of health and sickness
> in the economic relations of production and social relations of
> class and seeks preventions through changing the political
> relations of power.
>
> (Porter and Porter 1988: 102)

Given this promising start for a social medicine, based on the idea of a multicausal model of disease, which in turn implied an inter-disciplinary approach to medical intervention and medical training, how might we explain the growing dominance of a specialized medicine in the late nineteenth century, based upon fee-for-service, and the rise of a professional medical group with monopoly over allopathic medicine?

In the mid-nineteenth century, the medical profession was demor-alized and lacked effective professional regulation, organization and status. It could not demonstrate any significant therapeutic efficacy and it did not possess a monopoly over the delivery of medical services to any specific clientele. Because the whole system of the metropolitan general hospital had not been developed, the majority of patients received medical care in their homes on a private basis. Between 1875 and 1920, however, the status of general primary care was greatly transformed and the social standing of the general practi-tioner was significantly enhanced (Rosen 1983; Starr 1982; Starr and Immergut 1987). The growth in the demand for medical services was an effect of economic development, significant urbanization and the evolution of an urban system of mass transport. The dominance and the autonomy of the medical profession were rein-forced in this period by the growth of licensing laws which had the support of the state. A middle-class clientele developed with a specific demand for privatized scientific medicine, and furthermore the growth of an ideology of science greatly contributed to the recep-tivity of the population to technological medicine. In North America the combination of liberalism and individualism fostered the professional individualism of the doctor–client relationship which in turn was opposed to the social interventionism required by social medicine. Alongside these cultural and social conditions, there were a number of major advances in medical technology and scientific knowledge which made surgery, treatment and hospitalization increasingly safe and effective. There were improvements in anaes-thesia, there was the important development of germ theory through the research of Semmelweiss, Lister and Pasteur, and there were major advances in antiseptic procedures following Lister's use of antiseptic precautions for surgery which were widely accepted by the 1870s. Similar developments took place in the evolution of scien-tific medicine in Victorian Britain (Youngson 1979).

There is a tension between scientific and social medicine, because the former developed on the basis of a privatized relation between doctor and patient to the exclusion of other professional intervention, and was based upon a monocausal view of disease grounded in the germ theory as the foundation of a medical model. By contrast, social medicine implied the development of an inter-disciplinary approach to public illness based upon state interven-tion in the management and regulation of the environment rather than the medical management of the patient. The symbolic arrival of scientific medicine to social dominance was signalled by the publication of the Flexner Report in 1910 by Abraham Flexner, who proclaimed the importance of scientific medicine and provided a model for future medical development and medical training, not only in North America but in Europe, in his *Report on Medical Educa-tion in the United States and Canada*.

According to the Flexner Report, scientific medicine would require an extensive and protracted university-based training in scientific medicine and the implication of this requirement was that medical practitioners would come only from the middle and upper classes, because the cost of medical training would be, over a long period of time, quite prohibitive. Following the Flexner Report, 'the necessity for a college degree and the four year curriculum allowed only upper class students to continue to study medicine' (Berliner 1984:35). The report also had the effect of reducing the admission of blacks and women into the medical profession; for example, all of the five existing medical schools which provided a medical education for women to become physicians were closed. The recruitment of blacks and women into professional medical education did not show any signs of revival until after 1970 (Mumford 1983:322). The Flexner Report also both recognized and legitimated the dominance of a research-oriented scientific medicine, in which the biological sciences and laboratory training were to provide the foundation of medical education as a whole. While the dominance of a research-oriented scientific medicine looks like the dominance of allopathy over homoeopathy, Berliner (1984:35–6) makes the important point that in its early period 'scien-tific medicine was a unique way of organising the clinical experience, but it was not, at that time, a clinical medicine . . . Since the research program of scientific medicine had not yet produced a significant number of clinically effective outcomes, public support for scientific medicine was based on the success of science in endeavours other than medicine'. Scientific medicine also involved an increasing specialization of knowledge and a division of labour often organ-ized around separate organs of the body rather than around an understanding of the whole person. There was also further sub-discipline specialization, for example molecular biology from biology. For Flexner, the good physician was someone with a specialized knowledge of the medical sciences and those disciplines that were relevant to the theoretical basis of medicine, and, second, the good physician was highly critical in the treatment of the evidence of experience; the production of these scientific doctors required a special type of medical training and specialization which were in practice to be modelled upon the curriculum of the Johns Hopkins Medical School (King 1982:299).

This specialization of medical knowledge also resulted ultimately in the spatial and functional separation of the medical faculty from other faculties of the university which had the effect of further reinforcing the professional isolation of scientific medicine from other disciplines. The geographical isolation of the medical school, while contributing to the social solidarity of medical students, clearly makes interdisciplinary scholarly work extremely difficult to achieve. Even within the medical faculties, basic science facilities are often removed from the clinical disciplines which tend often to be centred in hospitals and their clinics. This sub-specialization and its geographical isolation are further intensified by the technical character of medical scientific language and the rapid expansion of knowledge in physiology and pharmacology (Perrin and Perrin 1984).

It is common in the history of medicine to argue that the Golden Age of scientific medicine was located in the period 1910 to 1950 in which Flexnerian medicine was never significantly challenged; this period was also one in which the general metropolitan hospital came to dominate the health-care system, as that location within which scientific medical practice had its primary focus. The growing importance of the general hospital was clearly associated with the growing status and prestige of the scientifically trained professional general practitioner within the community. There were in addition significant developments in the training of nurses for a specific place within the medical division of labour. Improvements in hygiene and sanitation within hospitals also had the consequence of significantly reducing high morbidity rates, thereby making hospitals safe for a middle-class clientele who became the main audience for the new medical technology (Larson 1977). It was the great era of the medical-industrial complex (Ehrenreich and Ehrenreich 1970). A number of writers, especially through the influence of Paul Starr (1982), have suggested that the Golden Age of scientific medicine may either have terminated or been transformed by changes in the economic basis of health-care systems and more generally by the changing character of the corporation within the capitalist economy (Cockerham 1986; Navarro 1986; Starr and Immergut 1987).

It is possible to present a fairly broad context within which we can understand the erosion of research-centred, specialized scientific medicine. First, there has been an important change in the disease structure of contemporary society. There has been a shift from acute to chronic disease and illness. For example, the leading causes of death in the United States at the turn of the century were influenza, pneumonia, tuberculosis and gastroenteritis; in the 1980s the leading causes of death by contrast were diseases of the heart, malignant neoplasms, vascular lesions of the central nervous system and accidents (Turner 1987:8). These changes in the character of disease are related to the ageing of the population and the success of scientific medicine in providing solutions to the acute illnesses of the nineteenth century, especially for contagious disease. In the United States the population over the age of 65 in 1900 was 4 per cent, which had increased to over 11 per cent by 1980, but the projection for the year 2050 is for almost 22 per cent of the population to be over the age of 65 (Cockerham 1986:33). The elderly are more likely than any other age group in society to require hospitalization and medication, partly because minor diseases become more rapidly transformed into life-threatening conditions which can no longer be treated in the home; in short, the elderly make far greater demands on the health-care system than other age groups (Russell 1981). In summary, there has been a transition in mortality towards diseases which do not have a specific or exclusive biological origin, to mortality rates which are explicable in terms of degenerative diseases, accidents and suicides. Thus there has been a change towards diseases which have 'a strong social component and a multifactorial etiology e.g. cancer, heart-disease, cerosis and arteriosclerosis' (Berliner 1984:40).

The consequences of these changes for conventional medical training are dramatic. The changing character of disease and the growth of the dependent population require a change in the medical curriculum towards interdisciplinarity, because the scientific medical curriculum, with its emphasis on acute illness and heroic medicine, can no longer provide appropriate medical solutions to the changing character of mortality and morbidity. Of course, there has been little evidence so far of any dramatic change in the scientific medical curriculum, which has retained an emphasis on the dominance of the fundamental components of the natural science disciplines, to the exclusion of a systematic study of the psychological, sociological, economic, political and environmental causes of human illness (Berliner and Salmon 1980). In addition to this hiatus between scientific medicine and the actual requirements of contemporary health-care systems, there has been a growing social critique of scientific medicine, specifically with reference to iatrogenic disease and to so-called 'unnecessary surgery' (Illich 1976; Inglis 1982; Navarro 1976).

Alongside the growing critical awareness of the social limitations of scientific medicine, there has been a mounting critique (from both the left and the right) of hospital management, hospital costs and the alienating consequences of hospitalization. Mental hospitals in particular came under criticism as primary illustrations of total institutions (Goffman 1961). In more recent years in Britain, there have been extensive enquiries into the mismanagement of hospitals and the neglect, or indeed abuse, of patients especially the elderly (Martin 1984).

These changes have, so to speak, brought the question of inter-disciplinarity and social medicine back on the agenda for the reform and training of medical practice in the late-twentieth century. Part of my argument has been therefore that there is typically a tension between social medicine and scientific medicine, and between medical dominance and interdisciplinarity in the academic curriculum. I now wish to consider two responses to the perceived need for a change in the medical curriculum, namely the notion of a sociology of health and illness and the idea of interdisciplinary research centres as models for university curriculum development.

THE SOCIOLOGY OF HEALTH AND ILLNESS

The story of the transition from medical sociology to a critical sociology of health and illness is well known. Medical sociology is seen to have had a late and uncertain start moving from an applied sociology in medicine to a more critical sociology of medicine (Cockerham 1986). A major turning point in the history of medical sociology was the development of the notion of a sick role by Talcott Parsons in *The Social System* (1951). Although the sick role concept has been criticized, it did indicate the theoretical grounds for an interdisciplinary approach to the nature of illness by combining elements of Freudian psychoanalysis with the sociological analysis of roles and a comparative cultural understanding of the importance of values in structuring the nature and distribution of illness in industrial societies (Holton and Turner 1986). However, while Parsons made the sociological analysis of illness central to sociological theory in the 1950s, medical sociology developed in two directions, one an applied sociology (sociology in medicine) and a more critical sociology of medicine (Strauss 1957). Criticism of traditional approaches led eventually to the notion of a sociology of health and illness as a more independent, relevant and theoretically informed perspective in the 1970s.

Although the advent of sociology of health and illness was often seen to be an optimistic and possibly imperialistic phase in the development of medical sociology (Strong 1979), over a longer historical period we can see that the relationship between sociology and medicine has been ambiguous and conflictual. Within the hierarchy of the sciences, the scientific credentials of both medicine and sociology are relatively low, both lacking the precision and mathematization characteristic of sciences like physics. Furthermore, to some extent sociology and medicine compete for the same audience, and sociology has often sought clinical status as an applied science alongside medicine. It was Lewis Wirth (1931) who recognized the development of clinical sociology as an important addition to the management and understanding of behavioural problems (especially in children). Wirth noticed that the sociologist, who had been previously neglected by psychiatry, could be an important addition to the therapeutic team, by developing a 'cultural approach' to behavioural problems. However, clinical sociology has never been fully established at a professional level (Glassner and Freedman 1979). It is the case however that medical sociologists have been able to penetrate the medical establishment, but the relationship between sociology and scientific medicine remains ambiguous. For example, while medical sociology often adopted a defensive position in response to critical commentary from either the medical profession or the basic sciences in the biochemical field (Jeffreys 1978), the intrusion of sociology into the medical curriculum can also be seen as an aggressive intervention in the medical academy. Of course, when the implicit or explicit critique of the sociology of health and illness was combined with a feminist analysis of the function of medical dominance in the lives of women (Oakley 1980) or with a political economy critique of the place of medicine in contemporary capitalism, especially in a period of economic recession (Doyal 1979), then the prospect for a successful or fruitful interdisciplinary programme between the social and the medical sciences has been remote and untenable.

Although the sociology of health and illness often adopts a radical stance with respect to the medical profession, it is not clear that sociology by itself could effect significant changes in the medical curriculum. In the past, the radical transformation of medical practices and academic medicine has been the consequence of political disturbance, a crisis of legitimacy and a consequent change in the relationship between the state and the professions (Foucault 1973). For example, significant changes in the French medical curriculum appear to be more the consequence of political change such as the Napoleonic revolution, the Gaulist regime and the student revolution of the late 1960s (Herzlich 1982; Jamous and Peloille 1970; Weisz 1980). The prospects of an interdisciplinary approach to medical analysis and health care may be paradoxically enhanced by the current erosion of the professional dominance of the physician as a consequence of greater government involvement in the health-care system, as a consequence of a stronger consumer lobby among the wealthy and the middle class, and finally as a consequence of the growing dominance of corporate power in the health-delivery system, especially in the private market (Starr 1982). One immediate result of corporate control of the private sector is the fact that the professional doctor in fact becomes merely an employee:

> The great irony is that the opposition of the doctors and hospitals to public control of public programs set in motion entrepreneurial forces that may end up depriving both private doctors and local voluntary hospitals of their traditional autonomy.
>
> (Starr 1982: 445a)

The professional autonomy which characterized doctors in the great era of scientific medicine has been gradually replaced by subordination to corporate power or to the mediation of the state (Johnson 1972). Starr has argued that as health centres are translated into profit centres, there will be a new requirement in the training of the physician by various forms of corporate socialization into business practice and commercial arrangements. Commercialization also 'constitutes a threat to the idea that professional physicians possess their own distinct body of general systematic knowledge'

(Ritzer and Walczak 1988:13). The full implication of these changes for interdisciplinarity are clearly matters of speculation. However, other changes following from government intervention in the higher education curriculum are topics which have already drawn the attention of educational theorists. In the next section therefore I examine the implications of a research centre model of interdisciplinarity in a period of monetaristic control following current attempts to reduce the welfare budget.

THE RESEARCH CENTRE MODEL: THE MCDONALDIZATION OF MEDICINE (Ritzer and Walczak 1988)

Both the sociology of health and illness and social medicine implicitly required the development of an interdisciplinary approach to health issues; this commitment to interdisciplinarity was ultimately based upon a philosophical view of 'the whole person' as the focus of health care. Scientific medicine is limited because it is based on a narrow, specialized and technical view of the human body as a machine which responds in a determinate way to the therapies derived from clinical experience and basic research. By contrast, proposals for interdisciplinarity were therefore based primarily upon a philosophical view of the body, the person and social relationships. There is, however, a very different set of pressures bringing about a reduction of specialization and an increase in interdisciplinarity which is essentially economic, and which is based upon a criticism of fundamental science which is often seen to be remote from real social problems and issues, and which is furthermore too costly in terms of real outcomes. A number of universities in the United Kingdom (such as Bradford and Sussex) have created interdisciplinary undergraduate programmes, while during the period of Thatcherite deregulation various attempts were made to create research centres which are problem-oriented, programmes which break down disciplinary specialization and arrangements which give financial incentives to universities competing for research funding. In the Netherlands, government pressure for higher educational reform resulted in the creation of an interdisciplinary social science programme at the Rijksuniversiteit of Utrecht in the 1980s. In Germany more theoretically guided schemes, influenced by the work of Helmut Schelsky, have attempted to establish centres of interdisciplinary research, for example the Zentrum für Interdisziplinaire

Forschung at the University of Bielefeld (Kocka 1987). Other prestigious illustrations would include the institutes for advanced studies at Princeton and Stanford.

In the United Kingdom, the reform of natural science research has been heavily influenced by the Rothschild Report, which has done much to break down disciplinary isolation, to focus natural sciences on specific problems relevant to the national interest and to academic arrangements which maximize funding and profitability. These macro-changes in the organization of science have brought about important transformations of careers in natural sciences, involving a transition from reputational to organizational career models (Ziman 1987). Before the Rothschild reorganization, the ideal or the characteristic scientific career was reputational, based upon specialization in one clearly defined sub-field of a discipline, followed by promotion on the basis of a public reputation demonstrated by papers presented at scientific meetings, membership of professional organizations and specialist publications. By contrast, the economic climate of Reagonomics and Thatcherism, which have involved continuous underfunding and cuts in the educational programme, requires constant changes of academic fields, research topics and research technology, because scientists with organizational careers will be forced constantly to change their jobs, their research locations and as a result their domestic arrangements. The absence of long-terms contracts in research institutes or the absence of tenured positions within the academy force highly trained scientists within the monodisciplines to change career paths in order to compete within the new research centres and the R & D institutions, where academic promotion and credibility will probably depend less on the quality and length of publications and more on the flexibility of research approaches and the ability to raise funding in a competitive market-place. In the new competitive deregulated academic market-place, those scientists who are unable or unwilling to adopt the new institutional career patterns will be faced by the threat of retrenchment or redundancy, as more traditional scientific fields are destroyed either as a consequence of government policy or by the sheer speed of change within the natural science field. These career patterns contrast sharply with the more traditional structure of the career of the established scientist. Within a non-competitive academic market characterized by tenure, there is the Matthew Effect (in which success breeds success) and the notion of 'undue persistence' in research or academic fields where individuals continue to draw upon their existing academic investment (or intellectual capital) long after the period in which such knowledge had relevance to a given field. Under these reputational conditions, scientists will be rewarded for their expertise in a specifically and narrowly defined area and for their persistence in a given topic. Within this traditional structure, therefore, the normal scientist rarely migrates between academic fields but at the very most drifts between adjacent problems. There is therefore much about the conventional academic market-place which precludes Kuhnian revolutions in knowledge, because the entire structure of reputation contributes to inertia. Of course, to argue that the reward system of science tends to produce special-ization and concentration on the solution of specific issues within sub-fields is to take a particular view of the character of scientific innovation. By contrast, some sociologists of science have emphasized the importance of scientific migration between various branches of a discipline or sub-discipline (Mulkay 1975). However, while the creation of a new discipline or sub-discipline produces career opportunities and new outlets for publication, there is, as it were, an ageing process in science where fields become established and stabilized as the original innovations become consolidated and recognized. In normal science therefore persistence and concen-tration are rewarded by power and resources which tend to reinforce this pattern of specialization (Zuckerman 1988).

In Britain, while research is undertaken within a variety of institutional settings (universities, research councils, public-sector research institutions and private-sector research groups), in the deregulation of scientific activity in the 1970s and 1980s a common set of assumptions has come to dominate much of scientific research, with a special emphasis on relevance and urgency in which funding is more and more based upon a customer-contractor principle. One consequence of this post-Rothschild reform era is the growing importance of interdisciplinarity in research and development groups, which has challenged the traditional isolation and special-ization of a monodisciplinary training. Because R & D organiza-tions are typically funded to solve specific problems, they will recruit graduates with a broad-based training which emphasizes flexibility and versatility. If this type of research becomes increasingly predominant, then it will begin to have important implications for university training and for the maintenance of monodisciplinarity. This type of interdisciplinarity, which I have suggested is brought about by strong economic pressure on the university system from the state, is not based upon a coherent philosophical position (or indeed on any strong educational philosophy at all) recommending interdisciplinarity as a norm. In this institutional context interdisciplinarity is more an unintended consequence of economic necessities than a consciously selected epistemological goal. This form of development produces *ad hoc* and short-term alliances and coalitions between scientific sectors, rather than an elegant and coherent map of the sciences. It results in a fragmented and decentralized scientific landscape. In this respect a Thatcherite model of science may, again for unanticipated reasons, ironically come to resemble the type of scientific world implied by postmodernism.

A POSTMODERN MODEL OF SCIENCE

It is neither possible nor necessary to enter into the complicated debate as to the origins, character and significance of postmodernism (Hassan 1985). For the sake of argument I shall take postmodernism to include at least the following: (1) the argument that the great rational project of the seventeenth century has come to an end, creating a situation in which there is no longer a single coherent rationality, but rather a field of conflicting and competing notions of the rational: thus we live in a fragmented, diversified and decentralized discursive framework; (2) because we can no longer appeal to the court of a single rationality and single morality, the 'grand narratives' of previous epochs (Science, Reason, Enlightenment, Humanity) have collapsed into a pile of conflicting myths and stories; (3) the hierarchies within science, morality and aesthetics have simultaneously broken down, thereby obscuring the relationship between elite culture and mass culture; (4) because of the impact of consumerism on all aspects of intellectual life, the institutional division between the university intellectual and the leader of pop culture has also become blurred and ambiguous, with the result that intellectuals may just as well seek an audience within the global television circuit as within the global academic market-place; (5) there is therefore an associated transition in which the aesthetic and the moral are combined, just as there is a transition from the the discursive to the figural. At various levels of modern society, these changes have given rise in architecture, in literary criticism, in design and more recently in the social sciences to the notion that we live in a constructed metaphorical reality in which, in the absence of a unifying or authoritative metaphor, culture is merely a *mélange*. Although postmodernism itself is clearly a fashion, it does point to significant and enduring changes in the social status and function of the intellectual and therefore to the social function and character of the university (Bauman 1987).

It is not therefore surprising that one of the great points in the debate over postmodernism, namely Jean-François Lyotard's *La condition postmoderne* (1979), was specifically a discussion of knowledge and of the possible function of the university in a postmodern period. The relevance of this discussion to university life is also underlined by the fact that the study, *La condition postmoderne*, was originally commissioned at the request of the Conseil des Universités of the government of Quebec and was dedicated by Lyotard to the Institute Polytechnique de Philosophie of the Université de Paris VIII (Vincennes). Lyotard starts by embracing the conclusions of sociological research on the notions of postindustrial society (with its emphasis on the communications revolution, cybernetics, the spread of computerization, the growth of knowledge banks, and the associated dominance of the service sector and the university within the information society) which he combines with contemporary trends in epistemology (associated with the work of Paul Feyerabend) to produce a critique of Habermas's view of consensual legitimacy; the result is that postmodernism involves what Lyotard calls an incredulity towards the metanarratives of conventional science.

In *La condition postmoderne*, Lyotard recognized two narratives for the legitimation of knowledge in two separate and distinct models of the university. The first was derived from the reforms adopted by Napoleon for higher education in which the main function of the university is to produce the administrative and professional personnel and skills which are essential for the stability of the state. In the second model, which was taken from the idea of Wilhelm von Humboldt and the founding principles of the university of Berlin, the university exists to provide a moral training for the nation, namely to bring about a Bildung-effect. The metanarrative which was behind this model of the university involved notions about the emancipation of the people and the legitimation of the state through some general conception about idealism (Habermas 1987). With the technological and computer revolutions of the postwar period, Lyotard argues that the metanarratives of legitimacy have broken down and the traditional divisions of labour and hierarchies within the university have equally disappeared. He thus argues:

> The classical dividing lines between the various fields of science are thus called into question – disciplines disappear, overlappings occur at the borders between sciences, and from these new territories are born. The speculative hierarchy of learning gives way to an immanent and, as it were, 'flat' network of areas of inquiry, the respective frontiers of which are in constant flux. The old 'faculties' splinter into institutes and foundations of all kinds, and the universities lose their function of speculative legitimation. Stripped of the responsibility for research (which was stifled by the speculative narrative), they limit themselves to the transmission of what is judged to be established knowledge, and through didactics they guarantee the replication of teachers rather than the production of researchers.
>
> (Lyotard 1979 (1986):39)

Because the universities are no longer committed to the production of ideals, they become merely instruments for the production of skills. At the level of epistemology and philosophy, the traditional questions about truth are similarly replaced by questions about pragmatics (that is reliability, efficiency and commercial value). Because monodisciplinarity is no longer necessarily the most efficient means of research or training in skills (and especially in fixed or permanent abilities), interdisciplinarity arises as the organization of knowledge relevant to the postmodern condition:

> The idea of an interdisciplinary approach is specific to the age of delegitimation and its hurried empiricism. The relation to knowledge is not articulated in terms of the realization of the life of the spirit or the emancipation of humanity, but in terms of the users of a complex conceptual and material machinery and those who benefit from its performers' capabilities.
>
> (Lyotard 1979 (1986):52)

Alongside these changes in the organization of knowledge, there is a greater emphasis on teamwork and disciplinary collaboration such that the traditional role of the single individual professor is destroyed, because the old individualism which legitimized the idea of the professor or the autonomous intellectual has given way to the ethos of the interdisciplinary team (with its problem orientation) and its computer banks and knowledge-storage capacities.

The commercialization of medicine and the translation of health into a calculation of profitability has been outlined by Paul Starr in his influential *The Social Transformation of American Medicine* (1982). For Starr, the medical idealism of the Hippocratic tradition has been translated into a marketable skill, but this particular case may in fact describe a very general commercialization of knowledge which will, along with the end of philosophy, announce the end of the university. The independent and autonomous general practitioner will become as archaic as the individual professor of a monodiscipline. The paradox is that the very success of the information revolution may have undermined the traditional status and function of the intellectual as 'a man of ideas'. For example:

> It was the intellectuals who impressed upon the once incredulous population the need for education and the value of information. Here as well their success turns into their downfall. The market is only too eager to satisfy the need and supply the value. With the DIY (Electronic) technology to offer, the market will reap the rich crop of the popular belief that education is human duty and (any) information is useful.
>
> (Bauman 1988:225)

The commercialization of intellectual life, alongside the commercialization of medicine as a specific instance, raises questions about the traditional institutions of professional knowledge, namely the licence and mandate to practise (Hughes 1958). Of course, the principal ideologue of free market competition has argued that 'licensure should be eliminated as a requirement for the practice of medicine' (Friedman 1962:158). If deregulation and postmodern epistemologies are both effects of changes in consumption, economic production and advanced technology, then we may expect the hierarchical division between scientific medicine and alternative medicine (like the distinction between high culture and mass culture) to collapse as the traditional autonomy of the medical profession is eroded through the invasion of corporations into the health market. Interdisciplinarity would then become not only a feature of the research institute and the training of medical personnel, but also a feature of consumption and production. The medical market would become a deregulated supermarket of health products just as the cultural world is, according to postmodern theory, itself a deregulated arena of hyper-consumption. There is therefore a peculiar (and to my knowledge unanalysed) relationship between the McDonaldization of culture, postmodernism and the dominance of Thatcherism as an economic and political principle.

CONCLUSION

I have considered four examples in medical history which have either promoted some form of interdisciplinarity as the ideal of medical training and intervention, or have implied interdisciplinarity as a desirable aspect of the medical curriculum. The first two (social medicine and the sociology of health and illness) were premised upon some notion of 'the whole person' set within a complex social environment where illness was the consequence of multiple causality (involving social, cultural and biological factors). Social medicine typically adopted a comprehensive and critical approach to the medical profession in which the physician would become merely one figure within a team of health police, whose aim was the complete regulation of society in the interests of a global condition of health. By contrast, the ambitions of sociology have been typically more modest; sociologists-in-medicine would be part of an inter-disciplinary approach to health care where sociology would be generally in a subordinate relationship to medicine and the natural sciences. Although social medicine and sociology were the product of rather different social and historical circumstances, they have tended to adopt a holistic perspective on medicine and have therefore at least implied a critique of the more specialized and narrow conception of the human being as a machine-like creature. The limited success of social medicine and sociology is at least partly an effect of the superior professional organization of medicine, which has until recently enjoyed the support of the state in protecting its licensed practice; secondly, the limitations of social medicine were partly the consequence of the success of scientific medicine in dealing with acute illness. The prospects for sociology may have been enhanced by the growing chronicity of illness and the more popular critique of scientific medicine.

By contrast, I have examined two other examples exerting pressures towards interdisciplinarity which probably share a similar cause, namely the commercialization of knowledge and professional practice. One type of interdisciplinarity (the creation of research centres based on teamwork with private and/or public funding) was the consequence of a political critique of the costs of modern technological medicine based upon monodisciplinary specialization.

This type of interdisciplinarity will further lead to a fragmentation of professional autonomy and transfer the traditional career path of the scientist from a reputational to an organizational career model. Secondly, the growth of the interdisciplinary research unit may be simply a small feature of a much larger historical and social transformation of modern societies by a process of postmodernization. If scientific medicine was simply the modern expression of the medical revolution of the seventeenth century based upon a Cartesian model of experimental science, then the challenge of postmodernism would deconstruct the metanarratives of medicine into fragmented and disorganized claims to power. Postmodernism exposes the fact that monodisciplines are federations of thematic components which are held together by the pressure of professional authority and the vested interests of their practitioners.

REFERENCES

Ackerknecht, E. (1953) *Rudolf Virchow, Doctor, Statesman, Anthropologist*, Wisconsin: Wisconsin University Press.

Abbott, A. (1988) *The System of Professions, an Essay on the Division of Expert Labour*, Chicago: Chicago University Press.

Bachelard, G. (1934) *Le nouvel esprit scientifique*, Paris: Alean.

Bauman, Z. (1987) *Legislators and interpretors: on Modernity, Postmodernity and the Intellectuals*, Cambridge: Polity Press.

Bauman, Z. (1988) 'Is there a postmodern sociology?', *Theory, Culture & Society* vol. 5:217–37.

Berliner, H.S. (1984) 'Scientific medicine since Flexner', pp. 30–56 in J.W. Salmon (ed.) *Alternative Medicines, Popular and Policy Perspectives*, London: Tavistock.

Berliner, H.S. and J.W. Salmon (1980) 'The holistic alternative to scientific medicine: history and analysis', *International Journal of Health Services* vol. 10:133–47.

Black, N., D. Boswell, A. Gray, S. Murphy and J. Popay (eds) (1984) *Health and Disease, a Reader*, Milton Keynes: Open University Press.

Bloom, S.W. (1988) 'Structure and ideology in medical education: an analysis of resistance to change', *Journal of Health and Social Behavior* vol. 29:294–306.

Bury, M.R. (1986) 'Social constructionism and the development of medical sociology', *Sociology of Health and Illness* vol. 8:137–69.

Charmaz, K. (1986) 'Social sciences in health studies: an interdisciplinary approach', *Sociology of Health and Illness* vol. 8:278–90.

Cockerham, W.C. (1986) *Medical Sociology*, Englewood Cliffs, New Jersey: Prentice-Hall.

Cohen, R.S. and T. Schnelle (eds) (1986) *Cognition and Fact, Materials on Ludwik Fleck*, Dordrecht: D. Reidel Publishers.

Doyal, L. (1979) *The Political Economy of Health*, London: Pluto Press.

Ehrenreich, B. and J. Ehrenreich (1970) *The American Health Empire, Power, Profits and Politics*, New York: Random House.

Featherstone, M. (1988) 'In pursuit of the postmodern, an introduction', *Theory, Culture & Society* vol. 5:195–215.

Foucault, M. (1973) *The Birth of the Clinic, an Archaeology of Medical Perception*, London: Tavistock.

Foucault, M. (1980) *Power/Knowledge, Selected Interviews and Other Writings 1972–1977* (edited by Colin Gordon), Brighton: Harvester Press.

Friedman, M. (1962) *Capitalism and Freedom*, Chicago: Chicago University Press.

Glassner, B. and J.A. Freedman (1979) *Clinical Sociology*, New York, Longman.

Goffman, E. (1961) *Asylums*, Harmondsworth: Penguin Books.

Gordon, J.S. (1984) 'Holistic health centres in the United States', pp. 229–51 in J.W. Salmon (ed.) *Alternative Medicines, Popular and Policy Perspectives*, New York: Tavistock.

Habermas, J. (1987) *Eine Art Schadensabwicklung*, Frankfurt: Suhrkamp.

Hassan, I. (1985) 'The culture of postmodernism', *Theory, Culture & Society* vol. 2:119–32.

Herzlich, C. (1982) 'The evolution of relations between French physicians and the state from 1890–1980', *Sociology of Health and Illness* vol. 4:241–53.

Holton, R.J. and B.S. Turner (1986) *Talcott Parsons on Economy and Society*, London: Routledge & Kegan Paul.

Hughes, E.C. (1958) *Men and their Work*, Glencoe, Ill: Free Press.

Illich, I. (1976) *Medical Nemesis*, New York: Pantheon.

Inglis, B. (1982) *The Diseases of Civilisation*, London: Hodder & Stoughton.

Jamous, H. and B. Peloille (1970) 'Changes in the French University-Hospital System', pp. 11–152 in J.A. Jackson (ed.) *Professions and Professionalization*, Cambridge: Cambridge University Press.

Jeffreys, M. (1978) 'Does medicine need sociology?', pp. 39–44 in D. Tuckett and J.M. Kaufert (eds) *Basic Readings in Medical Sociology*, London: Tavistock.

Johnson, T. (1972) *Professions and Power*, London: Macmillan.

King, L.S. (1982) *Medical Thinking, a Historical Preface*, Princeton, New Jersey: Princeton University Press.

Kocka, J. (ed.) (1987) *Interdiziplinaritat, Praxis, Herausforderung und Ideologie*, Frankfurt: Suhrkamp.

Larson, M.S. (1977) *The Rise of Professionalism, a Sociological Analysis*, Berkeley: California University Press.

Light, D.W. (1988) 'Toward a new sociology of medical education', *Journal of Health and Social Behavior* vol. 29:307–22.

Lyotard, J.F. (1979) *La condition postmoderne: rapport sur le savoir*, Paris, Les Editions de Minuit; English translation 1986, Manchester: Manchester University Press.

Martin, J.P. (1984) *Hospitals in Trouble*, Oxford: Basil Blackwell.

Mulkay, M.J. (1975) 'Three models of scientific development', *Sociological Review* vol. 23:509–26.

Mumford, E. (1983) *Medical Sociology, Patients, Providers and Policies*, New York: Random House.

Navarro, V. (1976) *Medicine under Capitalism*, New York: Prodist.

Navarro, V. (1986) *Crisis Health and Medicine, a Social Critique*, New York: Tavistock.

Oakley, A. (1980) *Women Confined*, Oxford: Martin Robertson.

Parsons, T. (1951) *The Social System*, London: Routledge & Kegan Paul.

Perrin, E.C. and J.M. Perrin (1984) 'Anti-intellectual trends and traditions in academic medicine', pp. 313-26 in W.A. Powell and R. Robbins (eds) *Conflict and Consensus, a festschrift in honor of Lewis A. Coser*, London: Collier Macmillan.

Piaget, J. (1970) *Main Trends in Interdisciplinary Research*, London: George Allen & Unwin.

Porter, D. and R. Porter (1988) 'What was social medicine? an historiographic essay', *Journal of Historical Sociology* vol. 1:90-106.

Ritzer, G. and D. Walczak (1988) 'Rationalization and the deprofessionalization of physicians', *Social Forces* vol. 67:1-22.

Rosen, G. (1979) 'The evolution of scientific medicine', pp. 23-50 in H.E. Freeman, S. Levine and L.G. Reeder (eds) *Handbook of Medical Sociology*, Englewood Cliffs, NJ: Prentice-Hall.

Rosen, G. (1983) *The Structure of American Medical Practice 1875-1941*, Philadelphia: University of Philadelphia Press.

Russell, C. (1981) *The Aging Experience*, Sydney: Allen & Unwin.

Sigerist, H.E. (1937) *Socialised Medicine in the Soviet Union*, London: Gollancz.

Starr, P. (1982) *The Social Transformation of American Medicine*, New York: Basic Books.

Starr, P. and E. Immergut (1987) 'Health care and the boundaries of politics', pp. 221-54 in Charles S. Maier (ed.) *Changing Boundaries of the Political*, Cambridge: Cambridge University Press.

Strauss, R.R. (1957) 'The nature and status of medical sociology', *The American Sociological Review* vol. 22:200-4.

Strong, P.M. (1979) 'Sociological imperialism and the profession of medicine', *Social Science and Medicine* vol. 13(A):199-215.

Strong, P.M. (1984) 'Viewpoint: the academic encirclement of medicine?', *Sociology of Health & Illness* vol. 6, 339-58.

Taylor, R. and A. Rieger (1984) 'Rudolf Virchow on the typhus epidemic in Upper Silesia: an introduction and translation', *Sociology of Health & Illness* vol. 6:201-17.

Turner, B.S. (1982) 'The government of the body: medical regimens and the rationalization of diet', *British Journal of Sociology* vol. 33:254-69.

Turner, B.S. (1985) 'The practices of rationality: Michel Foucault, medical history and sociological theory', pp. 193-213 in R. Fardon (ed.), *Power and Knowledge, Anthropological and Sociological Approaches*, Edinburgh: Scottish Academic Press.

Turner, B.S. (1987) *Medical Power and Social Knowledge*, London: Sage.

Virchow, R. (1848) *Die Mediszinische Reform, Eine Wochenschrift*, Berlin: Duck und verlag von. G. Reimer.

Weisz, G. (1980) 'Reform and conflict in French medical education 1870–1914', pp. 61–94 in R. Fox and G. Weisz (eds), *The Organization of Science and Technology in France 1808–1914*, Cambridge: Cambridge University Press.

Wirth, L. (1931) 'Clinical Sociology', *American Journal of Sociology* vol. 37, 49–66.

Youngson, A.J. (1979) *The Scientific Revolution in Victorian Medicine*, Canberra: Australian National University Press.

Ziman, J. (1987) *Knowing Everything about Nothing*, Cambridge: Cambridge University Press.

Zuckerman, H. (1988) 'The sociology of science', pp. 511–74 in Neil J. Smelser (ed.) *Handbook of Sociology*, Newbury Park: Sage.

Chapter 5

The body and medical sociology

INTRODUCTION: SOCIAL AND SOCIOLOGICAL PROBLEMS

The relationship between applied and theoretical science is deeply, and perhaps necessarily, problematic. The tensions between these two styles of scientific research are common to all aspects of both the natural and the social sciences (Ziman 1987). While in sociology the debate surrounding the virtues, or otherwise, of both applied and theoretical research is well established (Merton 1959), this issue has assumed a new urgency in the last two decades partly as a result of government pressure to force the social sciences to become more relevant, more applied, more cost-effective and more sensitive to the national needs of the economy. These pressures on the social sciences, especially in the United Kingdom, have given rise to a number of developments and experiments, which are designed to ensure the greater relevance of social science to what the government of the day defines as 'social problems'. For example, the growing interest in interdisciplinarity in the social sciences is, at least in part, a response to the crises in higher education and a response to government initiatives in higher education (Kocka 1987; Turner 1990). Although we can see some of these changes in the social sciences as effects of the crises in the Western capitalist economy, some features of interdisciplinarity may also be associated with the breakdown in high culture as a consequence of growing postmodern cultural conditions (Lyotard 1984). The attempt to make social science relevant at the cost of its theoretical integrity may be an attack on the cultural capital of the élites who run these disciplines.

In this chapter, however, I wish to focus attention on the specific problems of medical sociology, because the involvement of sociology in the medical field may turn out to be the classical test case of applied versus theoretical sociology. In this argument it is claimed that medical sociology simultaneously holds out the promise of successful cumulation of data and research in the applied area, while also offering possibilities for major theoretical advances. The challenge of our present context is, therefore, to marry these two potentialities to provide a more coherent theoretical body of knowledge, which is still relevant to the analysis and treatment of social problems. The primary and simple thesis of this discussion is that in order to survive the success of application, social science studies of health issues must have a clear and distinctive theoretical core, otherwise medical sociology will either be reduced to merely an administrative science, or fragmented and dissipated into a collection of incoherent lines of research.

Of course, medical sociology has already experienced a long and somewhat uncertain and chequered career. While Foucault (1980:151) has argued that sociology (or more narrowly medical sociology) had its origins in nineteenth-century social medicine (specifically in enquiries into the health status of the working classes of the large industrial cities), it is more conventional to suggest that medical sociology emerged in the health economics of the earlier twentieth century, was elaborated as an applied science as a consequence of research into the morale of American soldiers (Clausen 1987; Elinson 1985) and received its first systematic theoretical elaboration in Talcott Parsons's concept of the sick role (Parsons 1951). Medical sociology subsequently established itself as a successful branch of sociology, although anxiety has been repeatedly expressed as to its true theoretical and analytical status within the curriculum of sociology. It has become common, following Strauss (1957), to make a distinction between sociology in medicine and sociology of medicine. The sociologist in medicine is a scientist who works directly with medical professionals in studying the socio-cultural conditions that are relevant to the existence of illness such that the problems of sociology in medicine are primarily defined by professional groups outside sociology iself. The research problems of sociology in medicine are thus determined by the institutional and political needs of medical schools, teaching hospitals and other public health agencies, but in the long term the research goals of sociology in medicine are determined by the complex interaction between the state and the economy. By contrast the character of sociology of medicine is no different from any other core component of the sociological curriculum and it may be defined as 'research analysis of the medical environment from a sociological perspective' (Cockerham 1986:2). This distinction is not necessarily the most helpful characterization of the institutional development of the sociology of medicine, since for example it does not necessarily apply so readily to Europe (Claus 1983) and it has also been argued that in recent years this division has broken down because 'most research in medical sociology today, regardless of whether it is in a sociology department of a university or in a medical institution, deals with practical problems. In fact, many medical sociologists hold joint appointments in both settings' (Cockerham 1986:2). Although the relationship between applied and theoretical (or sociology in medicine and sociology of medicine) varies considerably across different societies (Cockerham 1986), many sociologists would nevertheless want to maintain a distinction between an applied medical sociology which was relatively uncritical, and a more theoretically directed sociology which is specifically critical of both the medical profession and the social causes of illness in advanced industrial societies.

Some indication of this tension may be derived from the emergence of the sociology of health and illness and the political economy of health and illness. As a critical reaction to the institutionalization of medical sociology as a branch of sociology serving powerful institutions, many sociologists welcomed the emergence of the sociology of health and illness, which specifically addressed the social causes of illness and disease, often from the patient's point of view rather than from the élite professional perspective. The publication in 1979 of the English medical sociology journal with the title *Sociology of Health and Illness* was perhaps indicative of a change in academic direction towards a more independent perspective on health issues. Alternative radical perspectives in the social sciences were signalled by the publication of Ivan Illich's *Medical Nemesis* (1975) and by the Marxist work of Vincente Navarro (1977; 1978; 1986). The framework of a radical political economy of health and health care is now a relatively well established aspect of the social science approach to health issues (McKinlay 1984).

Thus the demise of medical sociology has often been predicted. For example, Freidson in 1978 talked about the decline of medical sociology and anticipated that it might even vanish. These pessimistic diagnoses have been rejected with equal force (Elinson 1980; 1985) and it is also possible to point to convincing evidence of the vitality of medical sociology; for example, it is now the largest section of both the American Sociological Association and the British Sociological Association. Other commentators have optimistically noted that 'the extensive growth of sociological literature in academic medicine is further evidence of the rising status of the medical sociologist' (Cockerham 1986:13). However, these positive indicators of the consolidation of medical sociology may also disguise the continuity of more traditional problems. For example, while Parsons's concept of the sick role has now been subjected to extensive critical evaluation (Holton and Turner 1986; Turner 1987a), it is not clear that medical sociology currently has any specific integrating theme or powerful theoretical structure, which is able to give the field some coherence and direction. Parsons's analysis of sickness and the professions meant that in theoretical terms medical sociology was firmly locked into the major questions of sociology as such, namely the nature of social action (Turner 1987b), the production of social deviants and the institutions of social control. Within Parsonian sociology at least, medical sociology was forced to address itself to the profound, enduring and classical issues of the sociological imagination itself. By contrast, however promising and useful much contemporary medical sociology may be, it often fails to confront the fundamental questions either of sociology or of social science generally.

CRISIS IN MEDICINE, CRISIS IN SOCIOLOGY?

Thus from these introductory comments we can note that the tensions between applied and theoretical sociology have been endemic to theoretical sociology since its inception, and various solutions for this issue have been regularly suggested. In this section of my argument, I wish to show that there may be some new problems confronting sociology which are also closely related to the problems facing contemporary medicine. While the notion of crisis has often been overemployed (Holton 1987), there are changes occurring within medicine, and more broadly within the university system, which permit one to consider the current situation as one of crisis, and this crisis clearly has major implications for the future interaction between sociology and medicine. In a recent study of the system of professions, Abbott (1988) has argued that it is theoretically and empirically false to analyse professions in isolation, since we must conceptualize professions as a system in which there is a division of expert labour and an endless conflict over the legitimate jurisdiction of professional tasks. Therefore, insofar as sociology and medicine compete with each other for the explanation of illness, we cannot understand the position of the medical sociologist as a professional in isolation from other health-care occupations. Quite simply, changes in the nature of medicine (and the medical professions) have direct and systematic consequences for the practice of sociology either within the university or within the hospital.

For example, the changing structure of illness in the twentieth century has important implications for both professional medicine and sociology, because the change from the prominence of infectious to chronic degenerative illnesses has key significance for medical training and for the effectiveness of traditional therapeutic regimes. The rise of the medical profession to social dominance in the final decades of the nineteenth century was closely related to the success of germ theory, new practices in surgical intervention and the reliability of new drugs. The publication of the Flexner Report in 1910 was an important step in the institutionalization of scientific medicine in professional training (Berliner 1984).

By contrast, the relevance of medical sociology appears to be enhanced by the growing importance of chronicity in the character of disease in the twentieth century. The ageing of populations, the increasing importance of chronic illness, the impact of environmental changes on the disease structure, and growing public criticism of both the ineffective character of much contemporary allopathic medicine in fundamentally changing the current pattern of morbidity and mortality and its cost have created an environment within which social science appears to be able to offer an alternative perspective on long-term illnesses which are not amenable to conventional scientific medical intervention. Discontent with medical provision appears to be associated with the following factors: '(1) changes in the disease structure of modern societies; (2) changes in demographic patterns; (3) changes in the patient–physician relationship; (4) the limitations of the hospital; (5) problems associated with the technological approach to medicine; (6) the problems associated with the machine model orientation of scientific medicine; (7) the focus on cure over prevention in research and practice; and (8) the cost of medical care' (Berliner 1984:40).

These changes open up a niche within the medical system for the intervention of social science in general and for medical sociology in particular, because the geriatric illnesses which have a chronic character not only require significant social changes in the organization of medicine, but demand the modification of social behaviour on the part of both patients and health-care providers because, without greater lay involvement in illness management, there cannot be significant reductions in cost associated with increases in the effectiveness of medical intervention. In short, the decline in heroic medicine and the growth of what has been referred to as 'environmental illness' have created significant occupational opportunities for medical sociology.

While this may be an optimistic picture from the point of view of sociological involvement in health care, there are other changes in the organization of medical intervention which imply major, but negative, reorganization of both medicine and social sciences. The great global crises in the economy from the early 1970s, with the oil crisis and the permanence of the economic recession in the world system in the 1980s, brought about a number of major changes at government level, which were also associated with the growing dominance of the New Right. These changes have introduced a managerial climate in the provision of health care. The crippling cost of public provision within the welfare system led a number of governments to introduce more market-oriented monetaristic programmes to reduce the burden of medical provision, and to redirect responsibility for illness away from society towards the individual, who is now expected to engage in extensive preventive medical practices in order to prolong his or her longevity while also reducing morbidity. Examples of these changes include the redirection of the health-care system towards privatized insurance bases, the commodification of health-care provision, and the introduction of new managerial practices into the maintenance and organization of health-care systems (Paci 1987; Starr and Immergut 1987). In Britain, the result was the adoption by the Thatcher government of a new set of managerial criteria and priorities for restructuring the welfare system (Cousins 1987). Although in America the origins of New Right policies have been rather different, the consequences for health care have been somewhat similar (Pampel and Williamson 1988; Quadagno 1987; Rimlinger 1971).

The extensive privatization of scientific medicine in North America, the growing importance of commercial chains of hospitals, the super-profits of the pharmaceutical industry, the dominance of managerial norms of performance in health-care delivery and the general commodification of health have major implications for the professional standing of doctors who may be increasingly subordinated to the position of an employee of large health organizations (Starr 1982). These changes in the economic funding of high technology medicine are also connected with radical changes in the knowledge base of contemporary medicine such that there is a growing professional crisis in health education as a result of the revolution in biochemical understanding of disease processes. In America, corporations are also increasingly contributing to degree-granting programmes, which threaten the traditional alliance between the professional and the universities. These changes have 'been accompanied by serious attacks on the utility of university education' (Abbott 1988:211). Changes affecting the organization and status of the medical profession include

> Antitrust decisions, corporatization, conglomeration, bureaucrat-ization, technological change, unionization, and the rise of McDoctors (no appointment, walk-in medical facilities modelled after fast-food restaurants), third-party payers, HMOs (Health Maintenance Organizations), prospective payment systems based on pre-set DRGs (Diagnostic Related Groups), and the like.
>
> (Ritzer and Walczak 1988:1–2)

The implication of this emphasis on the managerial effectiveness and profitability of health-care delivery systems may lead to a massive McDonaldization and commercialization of American health care (Navarro 1986). In summary, these changes in the structure and content of medicine are associated with revolutions in the basic sciences which underpin medical practice; a growing anxiety about the economic cost of medical education and health-care provision; and a significant expansion of medical technology and technique, making doctors increasingly dependent upon a powerful technology. The result of these macro-changes is that

> The practice of medicine has become less attractive, medical students' debts have grown, applications to medical schools have dropped sharply, budgets for medical education are being curtailed for the first time in decades, and a large part of the lay culture is moving towards wellness. At issue is the

relationship of medical education as a social institution to the
health care system and to the training experience of future physicians who will take care of patients.

(Light 1988:307)

In the past, a number of conventional medical sociologists have
suggested, at least by implication, that, given the institutional and
intellectual problems which have confronted especially British
sociology in the postwar period, the salvation of sociology would
be to move into the medical faculty. This shift in institutional
geography would thereby indicate the transformation of the
professional status of the sociologist by associating sociology with
real science. However, recent changes in the practice and organization of medicine indicate that this geographical solution will not
in itself be sufficient to create an enduring professional basis for
a scientific sociology of medicine, because the commercialization
of academic medicine will bring about an erosion of the professional status of medical practice itself.

Thus, the problems of applied sociology are well known. An
applied sociology would involve the direction and regulation of the
discipline by external bodies so that sociological problems would
be imposed externally. There would consequently be little or no
systematic funding for basic research and for theoretical elaboration and development. The net result would be the fragmentation
of the discipline and its erosion within the university curriculum.
One additional development might be the fragmentation of
sociology, its dispersal and its reallocation within a variety of other
disciplines or fields such as epidemiology, clinical psychology and
community medicine. To these traditional problems, we can now
add the implications of the commercialization of social-science
research alongside the commercialization of medicine, whereby
those components of social-science research which could be shown
to contribute directly or indirectly to economic profitability would
be funded, while other types of research in sociology would be
starved of economic support. These changes in contemporary
medicine clearly have major implications for medical training, but
they should also have significant consequences for medical sociology
training and indeed for the development of sociology as such. These
changing circumstances within the university system create an even
greater urgency for a theoretically sophisticated defence of medical
sociology as a core topic within the sociology curriculum as a whole.

AGENCY AND STRUCTURE AS TOPICS IN MEDICAL SOCIOLOGY

It is possible to illustrate the main burden of my argument by comparing the status of a sociology of religion and medical sociology in the history of sociology itself. Although it is possible to argue that religion (at least organized, formal religion) has declined in social significance in the industrial societies in the twentieth century, the sociology of religion has played, and to some extent continues to play, a pivotal role in the theoretical advance of sociology. By contrast it is possible to argue that we live in a reality which is entirely medicalized (Zola 1972). However, medical sociology has not, and may never play, a pivotal role in the theoretical progress of sociology as a scientific discipline. We can take some minor confirmation of this argument from the fact that the classical sociologists (Comte, Marx, Weber, Durkheim, Mannheim and Simmel) made no sociological contribution to the analysis of health issues, either narrowly or broadly defined, while all of them made major contributions to the sociological analysis of religion. Expressing this claim in a more contemporary framework, I would argue that if we take some of the major textbooks in contemporary medical sociology such as Cockerham's *Medical Sociology* (1986), Mechanic's *Medical Sociology* (1968), Mumford's *Medical Sociology* (1983), Susser and Watson's *Sociology in Medicine* (1962) or Tuckett's *An Introduction to Medical Sociology* (1976), we can notice immediately that they do not directly, self-consciously or specifically address themselves to what might be regarded as the enduring and central problems of sociological analysis.

Why is this the case? One answer to this riddle is that to some extent religion was a sort of conceptual offence to the positivistic assumptions of social science which sociology itself sought to address. That is, religion raised fundamental problems about the nature of rational action, the place of ritual in human behaviour and the role of religious values in the maintenance of social stability. The phenomena of religion were in this respect seen to be fundamental to what we may regard as at least two major foci of sociological analysis, namely the problem of social order and the character of social action (Turner 1983). By contrast, it appears to be the case, at least superficially, that medical sociology is driven either by more problem-centred issues (such as the social aetiology of depression in the famous studies by Brown and Harris 1978) or they are inspired by what we might call low-level theoretical problems such as the debate about medical professionalization, doctor–patient interaction or the famous studies of medical education by Merton *et al.* (1957) and Becker *et al* (1961). My purpose is not to criticize these studies, since they are rightly often held up as major contributions to the development of empirical sociology and they can clearly be defended in terms of grounded theory (Glaser and Strauss 1968) or as theories of the middle range (Merton 1957). However, it would be wrong to confuse my claim (that an adequate medical sociology has to be informed by theoretical issues which are central to the sociological enterprise and which therefore contribute to the development of sociology as an analytical science of society) with the idea that medical sociology should be part of Grand Theory. A theoretically adequate medical sociology is not one which is divorced from empirical issues or from the necessity for problem-solving, but it has to make a contribution to the development of sociology as such. It may be the case that the very success of medical sociology in penetrating the medical establishment has somewhat divorced it from the mainstream of contemporary sociology.

In presenting this problem, I have used the expression that it *appears* to be the case that medical sociology is to some extent an atheoretical mode of sociological enquiry. In reality, I believe that the sociological study of medical issues, health and illness, and the medical professions is a theory-rich area of study and that the theoretical development of medical sociology may only require a greater theoretical reflexivity on the part of medical sociologists rather than a major reorientation of the sub-discipline. There are a number of reasons for believing that medical issues are theory-rich, but at least one central feature of this richness is the philosophical and sociological question: what is disease? One can argue that this question in medical sociology (and more broadly within the philosophy of medicine) has the same analytical place and status as the question: what is religion? That is the ontological status of disease entities raises fundamentally the underlying problematic of sociology which is (to use an inadequate but shorthand term) the question of relativism. At its most interesting, medical sociology (and perhaps more specifically medical anthropology) constantly points to the fact that many of the categories, both of professional and everyday use, which refer to illness, sickness or disease are variable across time and space. Some of the most interesting work done recently in the medical sociology and medical anthropology area has been precisely into the relationship between culture and illness, especially in societies undergoing rapid and profound social change (Ohnuki-Tierney 1984). More specifically in the area of the sociology of knowledge, it is clear that the contemporary debate about the constructed character of illness is of very general and profound interest to sociology as a whole (Bury 1986; Bury 1987; Löwy 1988; Nicolson and McLaughlin 1987). Perhaps the current interest in, and partial rediscovery of, the work of Ludwik Fleck in the philosophy of science points directly to the fundamental connection between the basic analytical questions of sociology and those of medical science (Cohen and Schnelle 1986).

Although this development in the area of the sociology of knowledge of medical facts is of very general importance, we need a more systematic way of organizing the theoretical structure of medical sociology, and it is valuable to start at least with two issues which are of central importance to sociology as a whole, namely the questions of agency and structure. The problem of agency is particularly interesting in relation to the debate about what is a disease, partly because the scientific theory of disease entities within the medical model does not allow, as it were, any base for voluntarism in the causation of human disease (Caplan *et al.* 1981). However, it is also unavoidable for sociology. Insofar as we take sociology, in the terminology of Weber, to be the interpretative study of social action, then sociologists must be interested in the problem of voluntarism, the normative character of human action and the interpretative quality of interactional social relationships. The problem of voluntary action in relation to the categories of sickness, illness and disease raises very basic questions about the meaning of illness (Turner 1987b).

It has already been suggested that, while Parsons's theory of the sick role has been somewhat discredited by subsequent research, at least the Parsonian framework attempted to understand health and illness within the context of the debate about voluntary action, social values and the integration of social systems (Holton and Turner 1986). The sick-role concept in Parsonian sociology has to be seen as one component within a far broader conceptualization of medical sociology. It is possible to argue that, following Parsons, the sociology of health and illness would be concerned with, at the level of the individual, the phenomenology of illness experience, at the level of the social the cultural categories of sickness in terms of a sociology of norms and deviants, and, at the societal level, with health-care systems and the politics of health from a macro-analysis of the function of illness in social systems (Turner 1987a). This framework, which embraces the idea of human agency in the selection of sick roles and the idea of cultural and structural constraints of action in terms of the organization of values and institutions within a society, is powerful and effective in the organization of medical sociology, both in teaching and research. However, there are even deeper theoretical issues which lie behind the idea of voluntary action which we need to pursue in order to develop medical sociology as a core area within sociology as a whole. One reason why disease is of fundamental interest to sociology is that it raises the question of the human body in relation to the categories of nature and society on the one hand, and to action and intention on the other. As a conclusion to the argument, therefore, my aim is to focus on the body as an organizing principle not only of medical sociology, but of sociology as such. More specifically, a sociology of the body provides an important, and possibly innovative, bridge between medical sociology and the core components of contemporary sociology.

TOWARDS A SOCIOLOGY OF THE BODY

Although the social sciences can be regarded as sciences of action, they have been typically dominated by a narrow set of assumptions about cognitive rationality, purposefulness, utilitarian maximization, and goal orientation. Furthermore, although sociologists like Parsons attempted to criticize these positivistic and rationalistic theories of action by arguing that norms and values had to be included in what he referred to as 'the action frame of reference', the dominant vision of action in the social sciences has put an emphasis on purposefulness, and practical rationality. The argument here is that the contemporary interest in the sociology of the body can be seen as part of more general movement in social science, but particularly in sociology, which has attempted to come to terms with the embodiment of the human actor and hence with the relationship between emotionality and feeling in relation to purposeful activity. My assumption is that the sociology of the body not only provides an important focus within sociology as a whole in contemporary work, but offers medical sociology, or more specifically the sociology of health and illness, an opportunity to become the leading edge of contemporary sociological theory. This development in medical sociology is therefore part of a much wider critique of the Cartesian assumptions of classical social science by a range of new social movements in contemporary theory. There is, from a variety of perspectives, an anti-foundationalist objection to conventional Cartesian positivism; in addition, there are critical objections to the utilitarian assumptions of economic theory (Anderson *et al.* 1988). In general, these critical developments can be seen as an alternative to the rational assumptions of modernism (Habermas 1987).

Why is the body the emergent topic of contemporary sociology? The first aspect would be precisely the problems associated with the Cartesian paradigm as a framework for the explanation and understanding of human behaviour and social action. One illustration of this is a new interest in the sociology of emotions as a region of sociology somewhat unexplored within conventional approaches. However, I would see this interest in the sociology of emotion as an exploration of the issues raised by an enquiry into the implications for sociology of theories of human embodiment. While utilitarian assumptions in the sociology of action point to a rational-decision model of human action, the sociological interest in the body forces us to look at affect, emotion and feeling as social components of action. As Elias (1987:340) has noted, sociology typically works with a division between the natural and the social:

> Thus sociologists may see the body as a topic of interest. But the prevailing routines of analytical isolationism make it easy to treat the body as a topic of sociological research set apart from other topics, perhaps as the subject matter of a specialism. There does not seem any need to explore the links connecting aspects of humans perceived as body, with other aspects perhaps perceived as disembodied. On a larger scale, too, human sciences of this type tacitly work with the image of a split world. The division of the sciences into natural sciences and others not concerned with nature reveals itself as a symbolic manifestation of an ontological belief – of the belief in a factually existing division of the world.

Recent enquiries into the sociology of the body may be treated therefore as an attempt to overcome this ontological division (O'Neill 1985; Turner 1984). The attack on this division is also occurring in phenomenology and philosophy, and in certain aspects of psychology (Levin 1985; Levin 1988; Shapiro 1985).

The second context or social movement within which an interest appears to emerge is in relation to postmodernism, which can also be regarded as a critique of Cartesian rationalist grand narrative in the sphere of cultural representation. As a critique of a unitary notion of rationalism, postmodernism celebrates the figural, the emotional and the allegorical over conventional models of rationality. In this respect, the postmodern body is a typical feature of this critique of rational modernism (Boyne 1988; Lash 1988).

The third component in the resurrection of the body is the recent history of feminism which has also drawn attention to the negation of emotionality by masculine rationalism in the social science core; but feminism has also drawn attention to the problems surrounding the social constructions of the body, the problematic relationship between nature and culture, and the historical character of the male–female division. These feminist critiques have independently therefore drawn attention to the problematic absence of the body in the human and social sciences (Suleiman 1986). Although the feminist movement has had a very direct impact on medical sociology through, for example, the work of Ann Oakley (1980; 1984), feminist theories of the body have yet to have the full impact in the entire conceptual structure of the sociology of health and illness, and medical sociology. It is interesting that, while critical debates in postmodernism and feminism have been pushing social sciences towards a re-evaluation of embodiment, the body–nature division and the role of emotions and feelings in human action, Marxism has so far failed to come to terms with this challenge, and Marxist commentary on this arena is somewhat limited to the contributions of writers like Sebastian Timpanaro (1975).

The other changes which make a sociology of the body an urgent requirement of contemporary theory have to be located in the broader movements of modern culture. The condition for the contemporary turn towards the problems of embodiment may be located in consumerism and leisure. Within consumer culture the emphasis on body-beautiful and body-maintenance provides the conditions for expansions of the market for the sale of a new range of commodities related to the enhancement of individual prestige through bodily displays. The new leisure consumption and the body-beautiful culture stimulate a whole new market around hedonistic personal life-styles, making the body a target of advertising and consumer luxury. Thus, in the contemporary framework of hedonistic advertising, 'consumer-culture latches onto the prevalent self-preservationist perception of the body, which encourages the individual to adopt instrumental strategies to combat deterioration and decay (applauded too by state bureaucracies who seek to reduce health costs by educating the public against bodily neglect) and combines it with the notion that the body is a vehicle of pleasure and self-expression' (Featherstone 1982:18).

The new emphasis in body-beautiful culture on self-preservation and self-maintenance as part of a moral responsibility for our upkeep may also be closely associated with the ageing of the populations of the Western industrial societies; there is a constant change in the self-image of the aged with a new emphasis on activity, fitness and preventive medical care. The ageing of populations has brought chronicity to the forefront of medical problems, giving a special urgency to personal fitness. We can see that this broad range of changes in the Western industrial societies has provided conditions for a new emphasis on the body in culture, which to some extent is reflected in the interest in the body in the social sciences. It would be wrong, however, to see this new materialism as simply a departure from the religiosity of previous conceptions of the body. In fact, it is possible to argue that the new devotion to sport and fitness has been precisely a transference of asceticism from the religious to the secular arena. And this morality of the body is perhaps powerfully, if tragically, illustrated by the new fascination with AIDS and with AIDS as a metaphor of moral decay. Throughout human history the body has provided a rich treasure of metaphors for social order and stability, and has been an essential feature of human reflections on chaos and order (Douglas 1966). What is 'inside' and 'outside' the body provides a language for discussing what is inside and outside the social. AIDS merely represents the most recent version of the human debate on what is morally tolerable by reference to what is physically harmless. There is one further reason for the cultural prominence of the body in contemporary societies which is further illustrated by the AIDS debate. We are now familiar with the idea of the world as a global place, with the consequence that disease is a globalized issue.

I have suggested that this growing interest in the body provides a fruitful set of circumstances for the theoretical development of medical sociology since I argued that the, as it were, natural province of medical sociology is precisely the ambiguous status of the body in human cultures, the paradoxical relationship between nature, society and body, and the social role of illness in human cultures as a symbolic map of the political and social structure. Since medical sociology is ultimately concerned with the categories of illness, sickness and disease, the body is, so to speak, the absent core of medical sociology. However, it could be objected that, apart from some accidental convergence of interest, there is little reason to feel either optimistic or enthusiastic about this convergence as a way of developing medical sociology as a genuine analytical contribution to the broad development of sociology as a whole. In short, a sceptical critic could simply respond with the question: so what? In a number of previous publications (Turner 1984; Turner 1987a, 1987b; Stauth and Turner 1988) the importance of the body as an organizing principle of teaching and research in medical sociology has been partly outlined. Following the work of Foucault in *Discipline and Punish* (1979), we can treat the body as a target of practices of rationalities which seek to regulate and dominate the body of the individual and the body of populations (Turner 1985). The fruitfulness of this Foucault framework has already been demonstrated (Armstrong 1983) but the full import of this orientation is yet to be fully realized and acknowledged.

Until recently the history of medicine has been unfortunately somewhat divorced from medical sociology. The first answer to the question 'so what?' is that a sociology of the body provides, and has provided, an important point of convergence between a very broad Foucauldian interest in the body in relation to political surveillance, a new macro-historical analysis of medicine in relation to Western thought, a broad appreciation of the body as a metaphor of social relations and an integrating focus for the study of the relationship between medicine and sexuality. In addition to Foucault's own work on medical history (Foucault 1987), these new points of theoretical development are now having a broad impact on, for example, the history of psychiatry (Bynum *et al.* 1985). Of course, one might want to object therefore that, while the sociology of the body is an interesting focus of research, it broadly amounts to the assertion that the philosophy and phenomenology of the body may have a general impact on the future development of medical sociology and the sociology of health and illness by making it more philosophically sophisticated, more comparative in its view of history and more historically rooted in the self-reflexive understanding of the emergence of the social sciences.

However, I want to go well beyond such an assertion in response to the 'so-what?' question by further arguing that the sociology of the body also enables us to integrate into mainstream sociology of health and illness some general developments within existentialist philosophy and phenomenology. A sociology of the body permits us to perceive the theoretical relevance of key developments in phenomenological psychology and philosophy, in particular the work of David M. Levin (1985; 1988), but also enables us to see the work of Oliver Sacks within a somewhat more central perspective. Although Sacks's work on Parkinson's disease (1976), migraine (1981) and on becoming a patient (1986) have been for a long time regarded as imaginative classics in offering a new insight into the phenomenological experience of being ill, they have remained perhaps somewhat marginal to the central issues of the sociology of health and illness. They are often used as imaginative insights rather than as major texts. I would argue by contrast that the sociology of the body enables us more clearly to understand the relationship between illness as a loss of identity, the psychological transformation of personhood which often results from major illness, and the importance of body-image to well-being. In other words, the sociology of the body represents a major counter-position to the medical model and to reductionism in sociobiology because, in the concept of embodiment, we can break out of the dualism of the Cartesian legacy, phenomenologically appreciating the intimate and necessary relationship between my sense of myself, my awareness of the integrity of my body and experience of illness as not simply an attack on my instrumental body (*Körper*) but as a radical intrusion into my embodied selfhood. The sociology of the body consequently provides one of the few intellectually satisfying bases for an interdisciplinary approach to disease and illness as conditions of the total embodied person within a sociocultural context.

It may be that the sceptical critic does not feel entirely satisfied since I have so far suggested two reasons for the importance of the body. The first is that it provides an organizing framework for the study of genuine issues of a theoretical nature within medical sociology since a lot of research that passes for medical sociology is in fact occupational sociology, the sociology of professions, organizational sociology and economic sociology.

Secondly, I have argued that this framework establishes a genuinely interdisciplinary paradigm which enables us to integrate existentialist studies of disease with phenomenological appreciations of illness experience, and furthermore allows us to begin to deal with these issues within a properly comparative and historical framework. It may be that the 'so-what?' objection still remains in force, because I have yet to suggest some new lines of research and teaching which would be of considerable and possible innovative interest to medical sociology. I propose to offer two answers to this continuing objection. My claim is that a sociology of the body enables us to grasp the fact that the study of anorexia and obesity are not marginal concerns of a sociology of health and illness but define the major theoretical topics which embrace the complex relationships between cultural change, social structure, personal identity and body-transformations (Brumberg 1988). This theoretical perspective also enables us to see the underlying analytically parallel issues in the sociological analysis of amputation, ageing and anorexia, namely the impact on self-image of sickness and body change. In short, the sociology of the body provides an integrating theory which opens up systematically the common aspects of a great variety of human problems which are essentially grounded in our embodiment.

In addition, the sociology of embodiment enables us to outline and map out new areas of sociological enquiry which have been somewhat neglected in the past. At present we do not have a sociology of pain, although, if my argument about the importance of the sociology of embodiment is correct, sociology should be able to make an important analytic and therapeutic contribution to the understanding and management of pain. The complex and ambiguous relationship between the mind, the nervous system, the body and the experience of pain is now well established. In the late-nineteenth century, William James observed that in experiments with frogs it was possible to separate the brain from the rest of the body by making a section behind the base of the skull between the medulla oblongata and the spinal cord, a condition in which the frog continued to live but with very modified behavioural activities. By irritating the skin of the frog with an acid, James was able to observe some remarkable defensive movements:

> The back of the foot will rub the knee if that be attacked, whilst if the foot be cut away, the stump will make ineffectual movements, and then, in many frogs a pause will come,

as if for deliberation, succeeded by rapid passage of the opposite unmutilated foot to the acidulated spot.

(James 1890, vol. 1:15)

Although much progress has been made in the analysis of phantom-limb phenomena and the placebo effect, the analysis of pain is largely dominated by behavioural science approaches based on some gate-control assumptions in which the body is regarded as a machine or a system of communication. By and large, pain is treated as a problem within the sensory system, and very few researchers have followed up the suggestion of H.R. Marshall (1894) that we may regard pain as also an emotional condition involving affect and influencing the entire personality of the victim (Melzack and Wall 1982:215). If we recognize pain as an emotional state, then we are immediately considering the idea of the person as an embodied agent with strong affective, emotional and social responses to the state of being in pain. It is not necessary here to outline a systematic and detailed sociology of pain; the main point of this example is to draw attention to a neglected aspect of the sociology of health and illness for which a theory of embodiment is an essential prerequisite for understanding pain as an emotion within a social context.

CONCLUSION

In this argument I have criticized the theoretical underdevelopment of medical sociology and the sociology of health and illness. Second, I have claimed that some aspects of the crisis of sociology are directly related to, and indeed necessarily connected with, the changing status of medical practice in modern societies. For this reason, the importance of a theoretical defence of medical sociology becomes ever more urgent. It was then shown that the notion of agency and structure is an essential feature of theoretical sociology and that any understanding of the notions of health and illness must be set in the context of a discussion of these concepts. The next stage in my argument was to suggest that, in fact, the sociology of the body is the most important issue behind the question of agency and structure, since most social science theories of agency are rationalistic and cognitive, to the exclusion of affect, emotion and feeling on the part of an embodied social agent. In reply to the 'so-what?' criticism, I have suggested that the sociology of the body can be an organizing principle in medical sociology, that it provides a method for integrating existing approaches, that it creates an interdisciplinary basis for the integration of existentialist, phenomenological and sociological approaches, and finally it was argued that new areas of research could be opened up as a consequence to the notion of embodiment, and I have very briefly indicated the sociology of pain as one research focus.

Essentially the argument behind the sociology of the body is, first, that sociology is genuinely a sociology of action, and that the social actor is not a Cartesian subject divided into body and mind but an embodied actor whose practicality and knowledgeability involve precisely this embodiment. Clearly, this is not a new idea. For example, one crucial development in the emergence of the sociology of the body was an earlier tradition of philosophical anthropology which attempted to take into account radical changes in twentieth-century biology and evolutionism in order to provide a better social-science understanding of the human condition. The work of Arnold Gehlen was critical in this development (Gehlen 1988). In this respect the sociology of the body is not an idiosyncratic or particularly modern invention, but has its roots in phenomenological psychology and philosophical anthropology; both these positions can be further traced back to the work of Nietzsche on the will to power. I concluded my book *The Body and Society* with a quotation from Nietzsche and it is perhaps appropriate to finish this argument on a similar note. For Nietzsche the body was a more 'astonishing idea' than the idea of a soul.

REFERENCES

Abbott, A. (1988) *The System of the Professions, an Essay on the Division of Expert Labor*, Chicago: Chicago University Press.

Anderson, R.J., J.A. Hughes and W.W. Sharrock (1988) 'The methodology of Cartesian economics: some thoughts on the nature of economic theorizing', *The Journal of Interdisciplinary Economics*, vol. 2:307–20.

Armstrong, D. (1983) *Political Anatomy of the Body, Medical Knowledge in Britain in the Twentieth Century*, Cambridge: Cambridge University Press.

Becker, H.S., B. Geer, E.C. Hughes and A.M. Strauss (1961) *Boys in White, Student Culture in Medical School*, Chicago: Chicago University Press.

Berliner, H.S. (1984) 'Scientific medicine since Flexner', pp. 30–56 in J. Warren Salmon (ed.) *Alternative Medicines, Popular and Policy Perspectives*, London: Tavistock.

Boyne, R. (1988) 'The art of the body in the discourse of postmodernity', *Theory, Culture & Society* vol. 5(2-3):527-42.

Brumberg, J.J. (1988) *Fasting Girls, the Emergence of Anorexia Nervosa as a Modern Disease*, Cambridge: Harvard University Press.

Brown, G.W. and T. Harris (1978) *Social Origins of Depression*, London: Tavistock.

Bury, M.R. (1986) 'Social constructionism and the development of medical sociology', *Sociology of Health & Illness* vol. 8(2):137-69.

Bury, M.R. (1987) 'Social constructionism and medical sociology: a rejoinder to Nicolson and McLaughlin', *Sociology of Health & Illness* vol. 9(4):439-41.

Bynum, W.F., R. Porter and M. Shepherd (eds) (1985) *The Anatomy of Madness, Essays in the History of Psychiatry; Volume 1: People and Ideas*, London and New York: Tavistock.

Caplan, A.L., H.T. Engelhardt and J.J. McCartney (eds) (1981) *Concepts of Health and Disease, Interdisciplinary Perspectives*, London: Addison-Wesley.

Claus, L.M. (1983) 'The development of medical sociology in Europe', *Social Science & Medicine* vol. 17:1591-7.

Clausen, A. (1987) 'Health and the life course: some personal observations', *Journal of Health and Social Behavior* vol. 28(4):337-44.

Cockerham, W.C. (1986) *Medical Sociology*, Englewood Cliffs, NJ: Prentice-Hall.

Cohen, R.S. and T. Schnelle (eds) (1986) *Cognition and Fact, Materials on Ludwik Fleck*, Dordrecht: D. Reidel.

Cousins, C. (1987) *Controlling Social Welfare, a Sociology of State Welfare, Work and Organization*, Sussex: Wheatsheaf Books.

Douglas, M. (1966) *Purity and Danger, an Analysis of Concepts of Pollution and Taboo*, London: Routledge & Kegan Paul.

Elias, N. (1987) 'On human beings and their emotions, a process-sociological essay', *Theory, Culture & Society* vol. 4(2-3):339-62.

Elinson, J. (1980) 'Medical sociology: theoretical underdevelopment and some opportunities', pp. 373-81 in H.M. Blalock (ed.) *Sociological Theory and Research, a Critical Approach*, New York: Fress Press.

Elinson, J. (1985) 'The end of medicine and the end of medical sociology', *Journal of Health and Social Behavior* vol. 26(4):268-75.

Featherstone, M. (1982) 'The body in consumer culture', *Theory, Culture & Society* vol. 1(2):18-33.

Foucault, M. (1979) *Discipline and Punish, the Birth of the Prison*, Harmondsworth: Penguin Books.

Foucault, M. (1980) *Power/Knowledge, Selected Interviews and Other Writings 1972-1977*, Brighton: Harvester Press.

Foucault, M. (1987) *The History of Sexuality; Volume 2: The Use of Pleasure*, Harmondsworth: Penguin Books.

Freidson, E. (1978) 'The development of design by accident', pp. 115-33 in R.H. Elling and M. Sokolowska (eds), *Medical Sociologists at Work*, New Brunswick, NJ: Transaction Books.

Gehlen, A. (1988) *Man, his Nature and Place in the World*, New York: Columbia University Press.

Glaser, B.G. and A.L. Strauss (1968) *The Discovery of Grounded Theory*,

Strategies in Qualitative Research, London: Weidenfeld and Nicolson.

Habermas, J. (1987) *The Philosophical Discourse of Modernity*, Cambridge: Polity Press.

Holton, R.J. (1987) 'The idea of crisis in modern society', *British Journal of Sociology* vol. 38:502-20.

Holton, R.J. and B.S. Turner (1986) *Talcott Parsons on Economy and Society*, London: Routledge & Kegan Paul.

Illich, I. (1975) *Medical Nemesis*, London: Calder and Boyers.

James, W. (1890) *The Principles of Psychology*, New York: Macmillan, 2 vols.

Kocka, J. (ed.) (1987) *Interdisziplinaritat, praxis-herausforderung-ideologie*, Frankfurt: Suhrkamp.

Lash, S. (1988) 'Discourse or figure? postmodernism as a "regime of signification"', *Theory, Culture & Society* vol. 5(2-3):311-36.

Levin, D.M. (1985) *The Body's Recollection of Being, Phenomenological Psychology and the Deconstruction of Nihilism*, London: Routledge & Kegan Paul.

Levin, D.M. (1988) *The Opening of Vision, Nihilism and the Postmodern Situation*, London: Routledge.

Light, D.W. (1988) 'Toward a new sociology of medical education', *Journal of Health and Social Behavior* vol. 29(4):307-22.

Löwy, I. (1988) 'Ludwik Fleck on the social construction of medical knowledge', *Sociology of Health & Illness* vol. 10(2):133-55.

Lyotard, J-F. (1984) *The Postmodern Condition: a Report on Knowledge*, Minnesota: University of Minnesota Press.

Marshall, H.R. (1894) *Pain, Pleasure and Aesthetics*, London.

McKinlay, J.B. (ed.) (1984) *Issues in the Political Economy of Health Care*, London: Tavistock.

Mechanic, D. (1968) *Medical Sociology*, New York: Free Press.

Melzack, R. and P. Wall (1982) *The Challenge of Pain*, Harmondsworth: Penguin.

Merton, R.K. (1957) *Social Theory and Social Structure*, New York: Free Press.

Merton, R.K. (1959) 'Notes on problem-finding in sociology', pp. ix-xxxiv in R.K. Merton, L. Broom and L.S. Cottrell (eds), *Sociology Today, Problems and Prospects*, New York: Harper & Row.

Merton, R.K., G.G. Reader and P.L. Kendall (eds) (1957) *The Student Physician*, Cambridge, MA: Harvard University Press.

Mumford, E. (1983) *Medical Sociology, Patients, Providers and Policies*, New York: Random House.

Navarro, V. (1978) *Class Struggle, The State and Medicine*, London: Martin Robertson.

Navarro, V. (1986) *Crisis, Health and Medicine, a Social Critique*, London: Tavistock.

Navarro, V. and D.M. Berman (eds) (1977) *Health and Work under Capitalism, an International Perspective*, Farmingdale, NY: Baywood.

Nicolson, M. and C. McLaughlin (1987) 'Social constructionism and medical sociology: a reply to M.R. Bury', *Sociology of Health & Illness* vol. 9(2):107-26.

Oakley, A. (1980) *Women Confined*, Oxford: Martin Robertson.

Oakley, A. (1984) *The Captured Womb, a History of the Medical Care of Pregnant Women*, Oxford: Basil Blackwell.

Ohnuki-Tierney, E. (1984) *Illness and Culture in Contemporary Japan, an Anthropological View*, Cambridge: Cambridge University Press.

O'Neill, J. (1985) *Five Bodies, the Human Shape of Modern Society*, Ithaca and London: Cornell University Press.

Paci, M. (1987) 'Long waves in the development of welfare systems', pp. 179–200 in Charles S. Maier (ed.) *Changing Boundaries of the Political*, Cambridge: Cambridge University Press.

Pampel, F.C. and J.B. Williamson (1988) 'Welfare spending in advanced industrial democracies 1950–1980', *American Journal of Sociology* vol. 93(6):1424–56.

Parsons, T. (1951) *The Social System*, London: Routledge & Kegan Paul.

Quadagno, J. (1987) 'Theories of the welfare state', *Annual Review of Sociology* vol. 14:109–28.

Rimlinger, G.V. (1971) *Welfare Policy and Industrialization in Europe, America and Russia*, New York: John Wiley.

Ritzer, G. and G. Walczak (1988) 'Rationalization and the deprofessionalization of physicians', *Social Forces* vol. 67(1):1–22.

Robertson, R. (1970) *The Sociological Interpretation of Religion*, Oxford: Basil Blackwell.

Sacks, O.W. (1976) *Awakenings*, Harmondsworth: Penguin Books.

Sacks, O. (1981) *Migraine*, London: Pan.

Sacks, O. (1986) *A Leg to Stand On*, London: Pan.

Shapiro, K.J.S. (1985) *Bodily Reflective Modes*, Durham: Duke University Press.

Starr, P. (1982) *The Social Transformation of American Medicine*, New York: Basic Books.

Starr, P. and E. Immergut (1987) 'Health care and the boundaries of politics', pp. 221–54 in Charles C. Maier (ed.) *Changing Boundaries of the Political*, Cambridge: Cambridge University Press.

Strauss, R.R. (1957) 'The nature and status of medical sociology', *The American Sociological Review* vol. 22:200–4.

Suleiman, S.R. (ed.) (1986) *The Female Body in Western Culture, Contemporary Perspectives*, Cambridge: Harvard University Press.

Susser, M.W. and W. Watson (1962) *Sociology in Medicine*, London: Oxford University Press.

Stauth, G. and B.S. Turner (1988) *Nietzsche's Dance, Resentment, Reciprocity and Resistance in Social Life*, Oxford: Basil Blackwell.

Timpanaro, S. (1975) *On Materialism*, London: NLB.

Tuckett, D. (ed.) (1976) *An Introduction to Medical Sociology*, London: Tavistock.

Turner, B.S. (1983) *Religion and Social Theory, a Materialist Perspective*, London: Heinemann Educational Books.

Turner, B.S. (1984) *The Body and Society, Explorations in Social Theory*, Oxford: Basil Blackwell.

Turner, B.S. (1985) 'The practices of rationality; Michel Foucault, medical history and sociological theory', pp. 193–213 in R. Fardon (ed.) *Power and Knowledge, Anthropological and Sociological Approaches*, Edinburgh: Scottish Academic Press.

Turner, B.S. (1987a) *Medical Power and Social Knowledge*, London: Sage.

Turner, B.S. (1987b) 'Agency and structure in the sociology of sickness', *Psychiatric Medicine, Illness Behavior* vol. 5(1):29–37.

Turner, B.S. (1990) 'The interdisciplinary curriculum, from social medicine to postmodernism', *Sociology of Health and Illness* vol. 12(1):1–23.

Ziman, J. (1987) *Knowing Everything about Nothing, Specialization and Change in Research Careers*, Cambridge: Cambridge University Press.

Zola, I.K. (1972) 'Medicine as an institution of social control: the medicalizing of society', *The Sociological Review* vol. 20(4):487–504.

Part III

Regimes of regulation

Chapter 6

The government of the body
Medical regimens and the rationalization of diet

The rationalization of culture and social institutions has been a major theme of sociological thought, providing a connecting thread between Weber on disenchantment, Marx's theory of alienation, Lukács on reification and the analysis of forms of rationality by the Frankfurt School.[1] In more recent years, Michel Foucault's treatment of the growth of discipline through the systemization of knowledge in the form of examinations, timetables, registers and taxonomies has once more brought the question of the codification of discourse to the centre of sociological theory.[2] The formalization of thought and conduct has become a crucial topic in a wide variety of historical and sociological approaches to the constitutive features of Western society.[3] The norms of calculation, prediction and organization are thus regarded as fundamental to social arrangements of public and private life in industrial capitalism. The extension of the principles of rational calculability has been examined by sociologists in a diversity of social contexts – law, industry, education, science and so forth. Although Foucault has drawn attention to the development of medical discourse in *Madness and Civilization*[4] and in *The Birth of the Clinic*,[5] the problem of the formalization of medical thought, the discipline of the body and the organization of diet as an illustration of rationalization has been generally neglected in sociology. The issue of formal medical knowledge can be appropriately raised in the context of a traditional sociological debate about the development of social classes in capitalist society through the theoretical perspectives of Weber and Foucault.

By way of introduction, one can begin by pointing to the parallel between religious asceticism and the medical regimen. The etymological relationship between 'regimen' and 'asceticism' is obvious enough. 'Regimen' is from *regere* or rule and refers, as a medical term, to any system of therapy prescribed by a doctor, especially a regulated diet. It also carries with it the archaic meaning of 'a system of government', but we might legitimately extend it to include 'the government of the body'. By comparison, the ascetic is someone who, again by a system of rules, practises self-discipline. 'Asceticism' comes from the Greek term for monk (*akētēs*) and from exercise (*askeō*), but it is also associated with the notion of working, or practising on, metal and thus with any disciplined practice; it can be claimed that asceticism and medical regimens are disciplines of the body by reference to rules, programmes or timetables. Empirically speaking, they both commonly focus on diet and involve a government of food. Since one of the classical debates in sociology has been concerned with an alleged relationship between the asceticism of Protestantism and the capitalist discipline of labour, one is prompted to ask the question – was there, historically, a parallel between the religious and medical disciplining of the body in the context of developing capitalism? It is certainly the case that Max Weber in *The Protestant Ethic and the Spirit of Capitalism*[6] alluded to such a possibility. The main argument of the Protestant Ethic thesis was that people do not 'by nature' want to earn more and more, but merely to reproduce their traditional conditions of existence. The advance of capitalism required the separation of the worker from the means of production and the subordination of immediate instinctual gratification. Appetite and sexuality became the principal threat to the religious vocation, but also to a more general rational control of the instinctual life. Weber noted, in passing, that the answer to religious doubts and sexual temptation was the same – 'a moderate vegetable diet and cold baths'. There was, moreover, a convergence between the 'hygienically oriented utilitarianism' of Benjamin Franklin and those 'modern physicians' who advocated moderation in sexual intercourse in the interests of good health.

Of course, the existence of a contradiction between man as part of nature and as part of society has been a recurrent theme in social philosophy. Most social contract theories have built in to their presuppositions some notion that society is achieved at some definite cost to the individual, namely a loss of freedom or happiness. In Malthusian demography, the sexual satisfactions of individuals had to be limited by moral means otherwise human populations would be limited by the disasters of war, famine and disease. The classic expression of this pessimistic conflict between social order and personal happiness was presented in Freudian metapsychology. Thus in *Civilization and its Discontents*,[7] Freud posed an especially sharp contrast between instinctual gratification and civilization. This Freudian conception of the body/society contrast became influential in critical theory, especially for Herbert Marcuse and Jürgen Habermas. In this chapter, however, I want to concentrate on the analyses of Michel Foucault because of his obvious relevance to the topic of medical regimens but also and incidentally to comment on the parallel between Weber's emphasis on 'rationalization' and Foucault's studies of scientific discourse.

Foucault is rarely specific about the purpose of his philosophy; indeed it would be somewhat inconsistent with his view of ideas and authorship for us to talk about 'purpose' in this context. We can perhaps initially outline Foucault's argument by examining what he writes against. He wants to criticize those accounts of the history of ideas which (1) assume some teleological development of rational knowledge which is progressive, continuous and liberal, (2) assume that the development of knowledge is directly associated with or the cause of improvements in the human condition, (3) assume that the growth of rational systematic knowledge is a certain index of an extension of fundamental political freedom or, by contrast, that rational knowledge and political terror are always empirically divorced. In opposition to these assumptions, Foucault's work stresses the discontinuities rather than the continuities of knowledge. Thus, in *Madness and Civilization*, it is noted that the official history of madness on the part of institutional psychiatry treats the history of madness as a steady progress away from terminological misdescription and inhuman treatment. The pejorative notion of 'madness' is replaced by the neutral term 'insanity', while the old mad-hut and ship of fools give way to asylums and eventually to the moral correction of Pinel and Tuke. Foucault counters this official view with the argument that 'insanity' is the product of a new psychiatric discourse and thus the history of madness is characterized by conceptual discontinuity (or in Althusserian language by 'epistemological rupture'). Furthermore, the moral correction of Tuke's Quaker retreat replaced the chains of the old asylums with the more powerful bondage of individual conscience, religious guilt and family authority. Knowledge is not continuous and is not separated from the exercise of power. *Madness and Civilization* and *Discipline and Punish* consequently share common themes; any power relation presupposes 'the correlative constitution of a field of knowledge'.[8] In particular, the 'birth of the prison' is conjoined with penology and criminology, the rise of the asylum with psychiatry, 'the birth of the clinic' with clinical medicine and the confessional with the casuistry of sex. The systematization of knowledge and the institutionalization of power have, as their object, the control of the body and the subordination of desire to reason. Thus, pedagogy, demography, penology, psychiatry and so forth represent, or point to, the emergence of the detailed, discplined control of the body in a matrix of social settings – the classroom, the prison, the hospital and the asylum. In *Discipline and Punish*, Foucault is specifically concerned with the growth of examinations, timetables, taxonomies, classifications and registers which provided the means for the detailed surveillance and disciplining of the body. Active, unrestrained bodies were thus rendered 'docile'. The discipline which had traditionally characterized the monastery was now extended to the factory, the school, the prison and the asylum.

It is important for the argument of this chapter to indicate, at this stage, a number of rather interesting parallels between Weber's view of 'rationalization' and Foucault's discussion of the power/knowledge relationship through a number of historical studies of the systematization of thought. Both Weber and Foucault have been profoundly influenced by Nietzsche in their mutual rejection of the assumption that rational knowledge is an unambiguous benefit to human existence and that the history of human societies is one of progress involving a struggle between liberty and despotism, reason and terror. Weber discusses the rise and development of rationalization in the context of theological systematization; the extension of the principles of calculation, prediction and reliability in scientific knowledge to all areas of life; the decline of magic and superstition; the growth of bureaucratic forms of organization in the military, state and industry. These themes in Weber's account of Western society would provide some general parallel to Foucault's interest in the development of systematic discourse. There are, however, two specific examples in Weber's sociology where he directly touches upon the growth of systems of classification as illustrations of rationalization, namely the systematization of the forms of musical notation in the West and the growth of double-entry book-keeping in business accountancy. When Foucault writes about the monastery as the earliest form of discipline and treats the monastery as a model for the discipline which eventually emerged in the factory, he very clearly reproduces Weber's argument (or an aspect of it) from the Protestant Ethic. While Weber was not explicitly concerned with the body/knowledge relationship that dominates the work of Foucault, he was nevertheless interested in ascetic practices and the development of a sober, disciplined and rational life-style in capitalism.

It is valuable to draw out this relationship between Weber and Foucault as a corrective to a theoretical deficiency in the latter. Weber was far more explicitly aware of the importance of the sociological context of discourse – the social origins of beliefs, the carriers of knowledge, the social groups who were most likely to develop rational knowledge and the audience to which ideologies are addressed. In Foucault, by contrast, the discourse appears to be almost sociologically disembodied. The general problem of the sociology of rationalization is that which is crucial to any sociology of knowledge, namely the relative autonomy of processes of thought from the interests of social groups. In Foucault's treatment of discourse, there is a pronounced reluctance to reduce systematic thought to interests, especially the economic interests of social groups, so that the growth of formal knowledge appears to be one which is imminent in discourse itself. Weber, by contrast, treats the historical development of systematic calculability as both generic to formal rationality and as the product of the social interests of intellectuals, professionals and urban social strata. There is in most of Weber's accounts of rationalization an argument about the 'elective affinity' between the logic of formal reasoning and the specific interests of social groups. The rise of legal rationality, theological systematization, formal musical notation and scientific reasoning was the outcome of both the logic embedded in these forms of discourse and the professional and class interests of lawyers, theologians, musicians and scientists. In these terms, it is thus possible to use Weberian sociology to make Foucault's purpose more explicit, namely how did the systematization of knowledge in the form of tables, taxonomies and registers relate to the general process of social rationalization in the context of an emergent capitalist society? Since in Foucault the central issue would be 'What forms of systematic discourse about the body were correlated with power relations within capitalism with special reference to the discipline of the body?', he has examined Bentham's panoptican system as a crucial development in the organization of bodies within the factory, the school and the prison, but oddly enough seems to have neglected what appears to be the obvious illustration, that is, the growth of a science of diet as expressed, in practice, in the form of the medical regimen. The application of science to the problem of the food intake relative to different social classes is a crucial illustration of the discipline of the body in the context of social class relationships within capitalism.

METAPHORS OF THE BODY

Before coming to the principal illustration of dietary science, it is important to my argument to digress briefly on metaphors of the body and metaphors of society, because the growth of theories of diet appears to be closely connected with the development of the idea that the body is a machine, the input and output requirements of which can be precisely quantified mathematically. One of the earliest analogies for the body has been that between the political organization of society and the anatomy of the body – hence the 'body politic'. In Aristotle and in medieval writers, the structure and function of political institutions were typically compared with the organs and functions of the body. In the history of social philosophy biological metaphors for society have been equally persistent. Spencerian sociology and Social Darwinism are rather obvious illustrations, but it is interesting to recall the influence of H.L. Henderson's *The Fitness of the Environment*[9] and Walter Cannon's *The Wisdom of the Body*[10] on Talcott Parsons's early development of structural functionalism. In addition to the metaphor of politics, the human body has been conceived either as a work of art or as a machine. The development of mechanistic metaphors of the body seems decisive for the emergence of a scientific discourse of the body and the development of dietary classifications. Descartes's *Discourse on Method*[11] was an especially important turning point for the mind/body problem and the elaboration of iatromathematics. In Descartes's philosophy, the body, not requiring a soul, can function like a machine according to mechanical laws. The problem with the metaphor is that, whereas machines such as water-pumps, electric kettles and typewriters are built for a specific purpose, what is the purpose of the soulless human machine? While Descartes in rejecting final causes could not answer the question,

> the metaphor assumes significance immediately if applied to the concept of man who has a rational soul totally distinct from his

body. For the purpose of man's conscious and purposeful life, the body can indeed be considered as a machine that will run according to the manipulations of the machinist. And it will run all the better if it has no purpose of its own, if it is stripped of teleological assumptions and of the vegetative and animal soul with which the ancients endowed it.[12]

Descartes's machine metaphor laid the basis for seventeenth- and eighteenth-century medical rationalism especially through the work of Herman Boerhaave (1668–1738) and the international influence of the medical school of Leyden. It was this iatromathematical tradition which provided George Cheyne with the theoretical basis to his popular dietary schema in the 1740s.

GEORGE CHEYNE, THE DIETARY REGIMEN AND THE DOMINANT CLASS

Cheyne was born in 1671 or 1673 at the Mains of Kellie in the parish of Methlick, Aberdeenshire, and appears to have been a student at Marischal College in 1688, receiving an honorary MD in 1740 from the university. In 1690 Cheyne became the tutor of John Kerr (later the first Duke of Roxburgh) at Floors Castle and, sometime after 1693, attended Edinburgh University to study medicine under Archibald Pitcairne (1652–1713), who had returned to Edinburgh from Leyden where he had been professor of physic. It was at Leyden that Pitcairne had been strongly influenced by the development of iatromathematics. Following the work of Bellini, Borelli, Boyle and Descartes, Cheyne wrote a defence of Pitcairne's ideas in a book called *The New Theory of Fevers* (1701) which went through six editions between 1702 and 1753. Cheyne was elected a Fellow of the Royal Society in London in 1701/2 and by 1724 Cheyne had become a popular physician among the élite of eighteenth-century London. Sometime after 1715, Cheyne also began to visit Bath and extended his medical practice to include people attending the spa. Cheyne included among his friends and patients Samuel Johnson, David Hume, John Wesley, Alexander Pope and Samuel Richardson, while his books were dedicated to an impressive section of the aristocracy – the Earl of Chesterfield, Lord Bateman, Sir Joseph Jekyll, the Earl of Huntingdon and the Duke of Roxburgh.[13]

While Cheyne wrote on iatromathematical topics and natural religion, the majority of his publications were accounts of and advertisements for his system of dieting which he held to be the basic remedy for mental and physical illness as well as the basis of long life. His major publications were *An Essay of Health and Long Life*,[14] *The English Malady*,[15] *An Essay on Regimen*,[16] and *The Natural Method of Cureing the Diseases of the Body*.[17] These books were frequently re-edited and translated into French, German and Italian. Cheyne's publications on diet reflect a particular crisis in his own life and a general problem characteristic of the eighteenth-century upper and middle classes, namely the problem of chronic obesity. Cheyne arrived in London in the great heyday of London taverns and coffee-houses in the reign of Queen Anne. In the fellowship of 'Bottle-Companions, the younger Gentry, and Free-Livers', Cheyne was 'able to eat lustily and swallow down much liquor' with the result that his weight rose dramatically to around 448 pounds. Cheyne fell into a deep depression, being diagnosed as suffering from 'English melancholy', and found great difficulty in walking. After attempting a variety of diets, Cheyne eventually settled to a regimen of milk and vegetables, regular exercise on horseback, little alcohol and regular sleep. His health improved, his weight was reduced and he died happily at the age of seventy in 1743.

In presenting Cheyne's medical ideas, it is convenient to organize this discussion around his metaphor of the body, the analysis of the cause of illness in the eighteenth century, his classification of food and the prescription of appropriate diets for respective social groups. Following the inspiration of Harvey's experimental approach to the circulation of blood, Cheyne declared that 'An animal Body is nothing but a Compages or contexture of pipes, an Hydraulic Machin, fill'd with a Liquor of such a Nature as was transfus'd into it by its Parents'.[18] This complex system of pumps, pipes and canals can only be satisfactorily maintained by the correct input of food and liquid, appropriate exercise and careful evacuation. Digestion is crucial to the proper functioning of this machine and thus Cheyne gave special attention to the question of quantity and quality of food and liquid. Medical practice was seen to be secondary to sensible dieting in servicing this hydraulic apparatus.[19] He observed that

> Art can do nothing but remove impediments, resolve Obstructions, cut off and tear away Excrescences and Superfluities, and

reduce Nature to its primitive Order: and this only can be done by a proper and specific Regimen in Quantity and Quality, by Air and Exercise, and by well judg'd and timeous Evacuations and preparing the morbid Juices for easier Elimination.[20]

The body's hydraulic system becomes strained and damaged when the internal juices and fluids are in poor condition. Thus, blood which, on examination, is 'fizy, liverish, with either too little serum tho' clear, or too much but muddy' has to be treated by a 'trimming Diet'. In general, the body as a machine requires surveillance under appropriate 'Diaetetick Management'.

The causes of the major forms of sickness in eighteenth-century society have to be, according to Cheyne, traced back to changes in eating habits. Cheyne's analysis appears in part to resemble certain themes from Rousseau in that human disease and mental stress are the products of civilization and economic progress. It was the expansion of trade and commerce which had brought new, exotic and rich food and liquor onto the market; these strong, spicy substances were playing havoc with the English digestion. Cheyne observed that

> Since our Wealth has increas'd, and our Navigation has been extended, we have ransack'd all the parts of the Globe to bring together its whole Stock of Materials for Riot, Luxury, and to provoke Excess. The Tables of the Rich and the Great (and indeed of all Ranks who can afford it) are furnish'd with Provisions of Delicacy, Number and Plenty, sufficient to provoke, and even gorge, the most large and Voluptuous Appetite.[21]

The result was that sickness was most common among

> the Rich, the Lazy, the Luxurious, and the Unactive, those who fare daintily and live voluptuously, those who are furnished with the rarest delicacies, the richest foods and the most generous wines, such as can provoke the Appetites, Senses and Passions in the most exquisite and voluptuous Manner.[22]

In addition to the growing variety and quantity of food available on the market, the other major contribution of civilization to disease and human misery was in overcrowding. Cheyne attributed much of the blame for human illness to

> the present Custom of Living, so much in great, populous, and over-grown Cities; London (where nervous Distempers are most frequent, outrageous, and unnatural) is, for ought I know the

greatest, most capacious, close, and populous City of the Globe, the infinite number of Fires, Sulphureous and Bituminous, the vast Expence of Tallow and foetid Oil in Candles and Lamps, under and above Ground, the Clouds of stinking Breaths, and Perspiration, not to mention the Ordure of so many diseas'd, both intelligent and Unintelligent Animals, the crowded Churches, Churchyards and Burying Places, with putrifying Bodies, the Sinks, Butcher-Houses, Stables, Dunghils &c and the necessary Stagnation, Fermentation, and Mixture of such Variety of all Kinds of Atoms, are more than sufficient to putrify, poison and infect the Air for twenty Miles round it, and which, in Time, must alter, weaken, and destroy the healthiest Constitutions of Animals of all Kinds.[23]

The diseases of civilization are, therefore, diseases of abundance, not scarcity. Overconsumption, overindulgence and overpopulation are treated as the basic causes of eighteenth-century 'Distempers'. If Cheyne was convinced about the disastrous effects of overeating, he was equally emphatic about the evil of drink. He noted that

The Benefits a Person who desires nothing but a clear Head and strong intellectual Faculties, would reap by religiously drinking nothing but Water, (tepid or cold as the Season is) while he is yet young, and tolerably healthy, well-educated, and of a sober honest Disposition, are innumerable.[24]

Again it was among the affluent who could afford strong wines, powerful spirits and exotic drinks that 'Gout, Stone, and Rheumatism, raging fevers, Pleurisies, Small Pox or Measles' were most common. Their passions were 'enraged into Quarrels, Murder and Blasphemy; their Juices are dried up; and their Solids scorch'd and shrivel'd'.[25] Indeed, Cheyne was prepared to regard 'ferment'd and distill'd liquors' as the sole cause 'of all or most of the painful and excruciating Distempers that afflict Mankind; It is to it alone that our Gouts, Stones, Cancers, Fevers, high Hysterics, Lunacy and Madness are principally owing'.[26] Cheyne's medical discourses were thus addressed to a population which was sedentary, urban and affluent, and within that population to an élite of aristocratic and professional men. To guard against the evils of civilization, Cheyne offered his dietary management which was based on a classification of food and moral prescription. Indeed, we could describe his whole

Table 1 Cheyne's recommendation's for food and digestion

Easy to digest	Less easy to digest
Spring vegetables	Pears
Asparagus	Apples
Strawberries	Peaches
Poultry	Nectarines
Hares	Cows
Sheep	Horses
Kids	Asses
Rabbets	
Whiting	Salmon
Perch	Eel
Trout	Turbot
Haddock	Carp
	Tench
Pullett	Duck
Turkey	Geese
Pheasant	Woodcock
	Snipe
Veal	Red deer
Lamb	Fallow deer

approach as a religio-medical prescription for sober living and regularity.

Cheyne's classification of foods attempted to specify those substances which are most easily digested and thereby fall within the scope of our 'concoctive powers'. We can represent Cheyne's recommendations for food and digestion in a dichotomous table (Table 1). To these specific recommendations, he provided a number of general guidelines. First, 'the larger and bigger the Vegetable or Animal is, in its kind, the stronger and the harder to digest is the Food made thereof'.[27] Second, those foods which are of a 'dry, fleshy, fibrous substance' are more easily concocted than fatty, glutinous substances. Third, flesh which is white in colour is generally kinder to the digestion than flesh 'inclining towards the flaming colours' and, finally, the milder the taste, the better since food with 'strong poignant aromatick and hot taste' is best avoided. It is interesting that Cheyne did not refer to food which was not native to the British Isles, although he was a violent critic of punch which he associated with the West Indies. He also objected to the effect of civilization on cuisine in terms of the preparation of food because it stimulated taste and appetite in ways which he regarded as unnatural.

Cheyne thus presented a contrast between man in a state of nature without the existence of culinary arts which unnaturally stimulate the appetite and urban, sedentary man exposed to the dangers of civilization where art and a plentiful supply of rich foods interrupted the normal processes of digestion. He thus asserted that 'When mankind was simple, plain, honest and frugal, there were few or no diseases. Temperance, Exercise, Hunting, Labour and Industry kept the Juices Sweet and the Solids brac'd'.[28] He therefore specifically advised against exciting the appetite through elaborate preparation and treatment of food in favour of consuming food in its raw state –

> the Inventions of luxury, to force an unnatural Appetite, and encrease the load, which Nature, without Incentives from ill Habits, and a vicious Palate, will of itself make more than sufficient for Health and long life, Abstinence and proper Evacuations, due Labour and Exercise, will always recover a decayed Appetite so long as there is any Strength and Fund in Nature to go upon.[29]

In his therapy, Cheyne adhered to the classical Greek medical doctrine of 'contrary medicine'. Since the diseases of civilization were the result of excess and inactivity, the remedy was to be one of temperance and exercise. In this contrary medicine, the regimen of diet was the crucial element –

> It is Diet alone, proper and specific Diet, in Quantity, Quality and Order, continued in till the Juices are sufficiently thinn'd, to make the Functions regular and easy, which is the sole universal Remedy, and the only Mean known to Art, or that an animal Machin, without being otherwise made than it is, can use with certain Benefit and Success, which can give Health, long life and Serenity.[30]

While Cheyne is best known for his regimen of 'a total rigid Milk and Vegetable Diet, with aqueous Liquors', his medical writing is possibly more interesting for its attempt to specify distinctive diets for different social categories and age groups. He distinguished between four types of diet in terms of severity: the common diet, the trimming diet, milk, seeds, fruit and vegetable diet, and a total strict milk and seed diet. He went on to categorize diets suitable for different age groups (Table 2).

These general prescriptions were then further specified according

Table 2 Cheyne's diets for different age groups

Age	Diet
0–15 years	'to persevere in a gradually increasing temperat Diet, without fermented Liquors'
15–30 years	'to be only temperat in animal Foods and fermented Liquors'
30–50 years	'after Fifty to give up animal Food suppers, and fermented Liquors'
60 + years	'to give up all animal Food; and then every Ten years after to lessen about a quarter of the Quantity of their vegetable Food; and thus gradually descend out of Life as they ascended into it'

to variations in the amount of exercise performed by people in different social categories or occupations. For example, a man of 'an ordinary Stature, following no laborious Employment, undue Plight, Health and Vigour' should consume the following quantity of food and liquid – '8 Ounces of Flesh Meat, 12 of Bread, or Vegetable Food, and about a Pint of Wine, or other generous Liquor in 24 Hours'.[31] These quantities are to be varied according to the nervous disposition of the individual, their age and occupation. Cheyne had a lot of advice for professional classes, those in sedentary occupations and for intellectuals. For the 'learned professions', he recommended regular use of his 'domestick purge' – a mixture of rhubarb, wormwood, nutmeg and orange peel – and again he was very conscious of the consequences of alcoholism in the performance of professional tasks – 'If any Man has eat or drank so much, as renders him unfit for the Duties and Studies of his profession (after an Hour's sitting quiet to carry on the Digestion), he has overdone'.[32] Once a proper, regular diet has been established, the professional man has only two further requirements for sound health – (1) 'A Vomit, that can work briskly, quickly and safely' by 'cleaning, squeezeing and compressing the knotted and tumified Glands of the Primae Viae';[33] (2) 'Great, frequent and continued Exercise, especially on Horseback'.[34] Diet, regular evacuation and exercise would not restore the state of nature from which civilized man had fallen, but they would at least partly counteract the ravages of abundance, overcrowding and sedentarization.

It would be tempting to regard Cheyne's dietary management

as an ascetic discipline having an 'elective affinity' with the needs of nascent capitalism in subordinating the body to the routine of industrial production. The consequences of dietary management would be similar to those of Methodism. Some social historians have, for example, regarded Methodism as a religious movement which had the consequence of producing a disciplined, sober and semi-literate work-force, perfectly tailored to the production requirements of factory life.[35] At one level, dietary management could work in the same direction of producing a sober and athletic population whose healthy bodies would not disrupt production as the result of illness following 'irrational' eating and drinking habits. As we have seen, however, Cheyne's diet was addressed to a class of people that was professional, sedentary, urban and engaged in mental activity. They were, furthermore, people who could afford to eat, ride horses and enjoy the luxury of a regular vomit. The popularity of Cheyne's ideas was set in the context of the London élite, and the clientele of coffee-houses and taverns. Cheyne's views on diet were clearly irrelevant to the eighteenth-century working class whose consumption was closely restricted by low wages and whose diet was based predominantly on cereals. Over half of the typical labourer's budget in the eighteenth century was allocated to cereals supplemented by a small quantity of potatoes and meat.[36]

There is some evidence that Cheyne's views reached a wider audience through the mediation of John Wesley. From reading Cheyne's medical works, Wesley[37] came, according to his Journal 'to eat sparingly and drink water' and employed Cheyne's views in writing his *Primitive Physick*.[38] There was an obvious attraction between Wesley's religious asceticism and Cheyne's view of the Christian importance of maintaining the body in good health through sober living, regular hours, exercise and temperance. Cheyne's views may, through Wesley's *Primitive Physick*, have reached an audience in the middle or lower middle class, but it is still very doubtful that the full medical regimen had any relevance for the working class. Cheyne's dietary management involved a disciplining of the aristocratic, not the labouring, body. The health, and hence the dietary practices, of the working class are likely to be of interest to a dominant class only under the following set of circumstances. First, the insanitary conditions of working-class districts in congested urban areas represents a threat to middle- and upper-class health with the spread of contagious diseases. Second, high levels of unemployment increase the burden of taxation on local authorities which are held responsible for maintaining workhouses and asylums under poor law legislation. Third, conscription into the military reveals extensive incapacity in the male population, through disease and sickness, thereby raising questions about the ability of a nation to defend itself in conditions of mass warfare. All three circumstances tended to converge in the period 1850–1939. Cholera epidemics in the nineteenth century increased awareness about the importance of ventilation, sanitation and water supplies among middle-class sanitarian reformers. The Crimean, Boer, First and Second World Wars revealed the inadequacy of military administration, hospital provision, medical supplies and general standards of health. Evidence of widespread disability among the working-class population at the time of the Boer War gave rise to the national efficiency movement which was aimed at promoting discipline and health through physical training, temperance, compulsory military service and the Boy Scout association.[39] This concern for the fitness of soldiers was part of a more general movement to improve health standards by improvements in diet. Charitable institutions which advocated regular medical inspections in schools and the provision of school meals reflected both political and philanthropic interest in the relationship between health and destitution.

The thrust towards a scientific analysis of diet came, however, more from the debates about urban poverty, labour efficiency and the economic burden of incarceration in prisons and asylums. A scientific interest in the measurement of poverty in relation to budget and diet among the working class found its classic statement in Charles Booth's *The Life and Labour of the People in London*[40] and B. Seebohm Rowntree's *Poverty, a Study of Town Life*.[41] Rowntree, in particular, drew upon existing studies of diet – such as W.O. Atwater's *Investigations on the Chemistry and Economy of Food*[42] and *Dietary Studies in New York City in 1895 and 1896*,[43] Robert Hutchinson's *Food and the Principles of Dietetics*,[44] and *A Study of the Diet of the Labouring Classes in Edinburgh*.[45] Rowntree was concerned to obtain accurate measures on the energy requirements in terms of calories for the average working man. The three conclusions of Rowntree's York study were that, the diet of the 'servant-keeping class' is in excess of that necessary for health, the food supply for the artisan class is satisfactory if there is no 'wasteful expenditure on drink', and the diet of 'the labouring classes, upon whom the bulk of the muscular work falls, and who form so large a proportion of the industrial population are seriously underfed'.[46]

In Rowntree's empirical social science we can detect a Quaker concern for sobriety and discipline alongside the metaphor of the body as a machine, whose efficiency can be, in principle, measured by reference to specific quantities of calories and protein. His research was, of course, located within a broad and diverse concern about the adoption of appropriate genetic policies by the state to guarantee the survival of a healthy, intelligent population – a concern which had its roots in Social Darwinism, eugenics and moral statistics. The theoretical development of demography, dietary science, biology and eugenics went hand in hand with a set of political and social anxieties about the effect of the working class, not only on parliamentary democracy and social order, but on the very biological bases of civilized society. The criminal, the soldier and the working man thus became objects of scientific discourse and of political practice. Contemporary anxieties about obesity and dieting, slimming and anorexia, eating and allergy are part of the extension of rational calculation over the body and the employment of science in the apparatus of social control. We can claim, therefore, that the dietary practices of the eighteenth-century professional classes have gradually percolated through the social system to embrace all social groups in a framework of organized eating, drinking and physical training.

CONCLUSION

As a rationalization of eating practices, the growth of dietary science provides a potentially interesting area for examining Foucault's attempt to indicate the subtle connections between the body, knowledge and power. From a sociological point of view, one difficulty with Foucault's attack on conventional histories of knowledge is that he does not attempt to provide precise social linkages between knowledge and interests. In part, this refusal stems from his rejection of any casual explanations involving economic or sociological determinism. In this discussion of diet, I have attempted to provide some general relationships between industrialization, social classes and the growth of a discourse of food. Dietary tables were typically aimed at forms of consumption which were regarded as 'irrational' threats to health, especially where overconsumption was associated with obesity and alcoholism. These dietary programmes were originally addressed to those social groups which were exposed to abundance – the aristocracy, merchants and the professional groups of the London taverns and clubs. The irrationality of untutored consumption was held to be incompatible with the exercise of professional duties because the order of the mind was damaged by the disorders of digestion. In their original form, these dietary schemata had a strong moral content since obesity was not only a sign of physiological abnormality but also of moral deviance. The frugality of man in the innocent state of nature contrasted sharply with the excesses of man in civilized society. The religious imagery of man fallen from grace was combined with Rousseau's man in society; the route out of this condition was one of asceticism and diet to restore the mind and body to health. It was not until the latter part of the nineteenth century that the science of diet became important in the economic management of prisons and the political management of society. The principles for the efficient government of prisons and asylums were quickly applied to the question of an effective, healthy working class supported on a minimum but adequate calorie intake. In so far as the extension of rational control over social classes corresponded to the growth of dietetics and social science, the social survey of working-class diet represents an important illustration of Foucault's analysis of knowledge/power and Weber's treatment of asceticism/capitalism.

NOTES

1   Karl Löwith, 'Weber's interpretation of the bourgeois-capitalist world in terms of the guiding principal of "rationalization"', pp. 101–22 in Dennis Wrong (ed.), *Max Weber*, New Jersey, Prentice-Hall, 1970.
2   Michel Foucault, *Discipline and Punish, the Birth of the Prison*, Harmondsworth, Penguin Books, 1979, p. 27.
3   Norbert Elias, *The Civilising Process*, Oxford, Basil Blackwell, 1978; Wolfgang Schluchter, *Die Entwicklung des okzidentalen Rationalismus*, Tübingen, Mohr, 1979.
4   Michel Foucault, *Madness and Civilization*, London, Tavistock, 1967.
5   Michel Foucault, *The Birth of the Clinic, an Archaeology of Medical Perception*, London, Tavistock, 1973.
6   Max Weber, *The Protestant Ethic and the Spirit of Capitalism*, London, Allen & Unwin, 1965.
7   Sigmund Freud, *Civilization and its Discontents*, New York, Doubleday, 1958.
8   Michel Foucault, *Discipline and Punish, the Birth of the Prison*, Harmondsworth, Penguin Books, 1979, p. 27.
9   H.L. Henderson, *The Fitness of the Environment*, New York, Macmillan Company, 1913.
10 Walter Cannon, *The Wisdom of the Body*, New York, Norton, 1932.
11 R. Descartes, 'Discours de la méthode', *Oeuvres de Descartes*, Paris, Adam & Tannery, 1902.

12   Owsei Temkin, *The Double Face of Janus and Other Essays in the History of Medicine*, Baltimore and London, Johns Hopkins University Press, 1977, p. 276.
13   Henry R. Viets, 'George Cheyne, 1673–1743', *Bulletin of the History of Medicine*, vol. 23, 1949, pp. 435–52; Charles F. Mullet (ed.) 'The letters of Doctor George Cheyne to Samuel Richardson (1733–1743)', *University of Missouri Studies*, vol. 18, 1943, pp. 7–132.
14   George Cheyne, *An Essay of Health and Long Life*, London, 1724.
15   George Cheyne, *The English Malady, or a Treatise of Nervous Diseases of all Kinds as Spleen, Vapours, Lowness of Spirits, hypochondriacal and hysterical distempers, etc.*, London, 1733.
16   George Cheyne, *An Essay on Regimen, Together with Five Discourses, Medical, Moral and Philosophical*, London, 1740.
17   George Cheyne, *The Natural Method of Cureing the Diseases of the Body and the Disorders of the Mind Depending on the Body*, London, 1742.
18   Cheyne, 1740, p. ii.
19   Harris L. Coulter, *Divided Legacy, a History of the Schism in Medical Thought*, Washington, Wehawken, 1977, vol. 2, pp. 110 ff.
20   Cheyne, 1740, op. cit., p. iv.
21   Cheyne, 1733, p. 49.
22   Ibid, p. 28.
23   Ibid., pp. 54–5.
24   Cheyne, 1740, op. cit., p. xxxiv.
25   Cheyne, 1724, op. cit., p. 44.
26   Cheyne, 1740, op. cit., p. lvii.
27   Cheyne, 1724, op. cit., p. 23.
28   Cheyne, 1733, op. cit., p. 174.
29   Cheyne, 1724, op. cit., p. 29.
30   Cheyne, 1740, op. cit., p. x.
31   Cheyne, 1724, op. cit., p. 34.
32   Ibid., p. 38.
33   Cheyne, 1740, op. cit., p. xlvii.
34   Ibid., p. xlviii.
35   E.P. Thompson, *The Making of The English Working Class*, Harmondsworth, Penguin Books, 1968; Bryan S. Turner and Michael Hill, 'Methodism and the pietist definition of politics: historical development and contemporary evidence', Michael Hill (ed.) *A Sociological Yearbook of Religion in Britain*, vol. 8, 1975, pp. 159–80; Nicholas Abercrombie, Stephen Hill and Bryan S. Turner, *The Dominant Ideology Thesis*, London, Allen & Unwin, 1980.
36   G.D.H. Cole and Raymond Postage, *The Common People, 1746–1946*, London, Methuen, 1961; Fernand Braudel, *Capitalism and Material Life 1400–1800*, London, Weidenfeld & Nicolson, 1973.
37   John Wesley, 'A letter to the Right Reverend the Lord Bishop of London' in Gerald R. Cragg (ed.) *The Works of John Wesley*, Oxford, Clarendon Press, 1975, vol. 11, p.345; John Telford (ed.) *The Letters of the Rev. John Wesley*, Epworth Press, 1931, vol. 1, p. 11.
38   John Wesley, *Primitive Physick or an Easy and Natural Method of Curing Most Diseases*, London, 1752.

39 Bentley B. Gilbert, *The Evolution of National Insurance in Great Britain, the origins of the Welfare State*, London, Michael Joseph, 1966.
40 Charles Booth, *The Life and Labour of the People in London*, London, Macmillan, 1903, 17 vols.
41 B. Seebohm Rowntree, *Poverty, a Study of Town Life*, London, Thomas Nelson, 1902.
42 W.O. Atwater, *Investigations on the Chemistry and Economy of Food*, US Department of Agriculture, Bulletin No. 21.
43 W.O. Atwater, *Dietary Studies in New York City in 1895 and 1896*, New York, Chas. D. Woods, n.d.
44 Robert Hutchinson, *Food and the Principles of Dietetics*, Edinburgh, n.d.
45 Dr Dunlop, *A Study of the Diet of the Labouring Classes in Edinburgh*, Edinburgh, Schulze, 1900.
46 Rowntree, op. cit., p. 28.

Chapter 7

The anatomy lesson
A note on the Merton thesis[1]

THE ETHIC OF WORLD MASTERY

In *The Protestant Ethic and the Spirit of Capitalism*, Max Weber outlined a theory of asceticism as an aspect of the more general drive for salvation in the Abrahamic faiths.[2] In particular, within the Protestant tradition the quest for salvational certainty produced an ethic of inner-worldly asceticism, whereby the faithful sought control and regulation of their entire life-world. This irrational quest for certainty in the face of an absent God produced a culture of regulation. We are now familiar with the argument that the unintended consequence of inner-worldly asceticism was the production of secular callings within the world, whereby success in business was eventually taken to be the hallmark of inner perfection. For example, at the end of the Protestant ethic thesis, Weber quoted approvingly from the life of John Wesley, who recommended that we should earn all we can, save all we can, and give all we can away. It was through these mechanisms that the Protestant ethic was finally converted into the spirit of capitalism; or at least the Calvinistic doctrines of asceticism had an elective affinity with the requirements of business practice. Puritanism liberated the secular calling (*Beruf*) from its negative evaluation within Catholicism which had given ethical priority to the *vita contemplativa* (Poggi 1983:60).

The importance of Robert K. Merton's original publication in *Osiris* (1938) was to extend the Weber thesis on the specific characteristics of capitalism to the origins and nature of science and technology in the seventeenth century. Merton noted that various aspects of Puritan activism were not only compatible with the emergence of natural science but actually acted as a spur to

the development of modern science based on instrumental rationality. For example, the Puritan critique of idleness led to a perspective
on science as a useful and fitting activity for men of God. More
importantly, Puritan tastes were highly compatible with both
utilitarianism and empiricism. For Merton 'the combination of
rationalism and empiricism which is so pronounced in the Puritan
ethic forms the essence of the spirit of modern science' (Merton
1970:92). Rational experimentation, being part of the revolt against
the conventionalism of the Aristotelian legacy, drove out passive
contemplation of reality, and was legitimized as a religious investigation of nature as itself a reflection of the divine spirit. In Merton's
account, we can see that science, like the calling in business, was
part of an attempt to regulate and control reality in the interests
of some broader religious purpose. Because compromise with
worldly existence was not tolerable, the world 'must be conquered
and controlled through direct action and this ascetic compulsion
was exercised in everyday life' (Merton 1970:99). While mysticism
might encourage either a flight from the world or its passive
contemplation, active inner-worldly asceticism required not only
the inspection of reality by rational means, but its positive and
absolute control and domination in the interests of the religious life.

It can be argued on the basis of Weber's comparative sociology
that we are warranted in extending the idea of inner-worldly
asceticism to a broader concept of an ethic of world mastery.
Therefore, following Weber's historical sociology, we can understand the social history of Western, rational culture as a set of
cultural variations on the interaction and relationship between the
mind and body, flesh and spirit, nature and culture (Turner 1984;
1987a). Weber's sociology of asceticism is a reflection on a
fundamental tension or contradiction within the Judaeo-Christian
tradition which opposes the life of the spirit to the irrational dangers
and temptations of the world. Within Christian theology, the
concept of 'the world' embraces the widest range of human issues
including luxuries, desire, pleasure, idleness and the enjoyment
of the body. The contradictions between the life of the spirit and
human embodiment is perhaps best signified by the notion of 'the
flesh' in Christian theology. The flesh is a comprehensive symbol
for the animality of Man, from which the Christian's soul must
seek either flight or domination.[3]

I have argued elsewhere (Turner 1987a:19) that within Western Christian culture there emerged three great institutional

arrangements or responses to the fleshliness of our existence, namely the arenas of religion, law and medicine. These institutional superstructures are organized responses to the spiritual problems of human embodiment and the need for cultural management of this embodiment. This argument in part follows Arnold Gehlen's theory (1988:378) that we may understand religion as a direction system (*Führungssystem*) which permits active adjustment to the world. Law, religion and medicine are institutional arrangements which are societal responses to the embodiment of persons in the world and the reciprocal relations in everyday life between such embodied persons, since human needs cannot be resolved or satisfied autonomously without sociation (Stauth and Turner 1988). Religion can be regarded as a collection of ritual practices and beliefs which aim to regulate and restrain the embodied person in the interests of some spiritual goal; the word *religio* itself indicates that religion is related to those regulations, rules and bonds which constrain and control the person in the interests of wider collective and transcendental goals. Law, especially the criminal law, is concerned with the aim of creating and maintaining a set of social contracts as the basis of civilized order. Finally, we can treat medicine as a powerful regulation and restraint of the human body, creating discourses of disease which monitor, while also in part constituting, various forms of deviance (Zola 1972). The model which is developed here follows as an extension of Weber's analysis of the diverse tensions between the Christian ethic of asceticism and secular life, especially insofar as that world was reproduced by the act of sexuality (Weber 1966).[4]

These attempts to regulate the body through a network of social institutions and cultural standards are the core component of the ethic of world mastery. The implication of this argument is furthermore that we can regard the ethic of world mastery as the basis of the project of modernity, which is the imposition of instrumental reason over nature, social relations and personality. In their study of the enlightenment, Theodor Adorno and Max Horkheimer argued that European civilization had two separate histories, namely 'A well known, written history and an underground history. The latter consists in the fate of the human instinct and passions, which are displaced and distorted by civilization' (Adorno and Horkheimer 1979:231). Civilization is brought about by the denial of the human emotions which come to be defined as irrational. The subordination of the flesh required an intellectualization of life through the development of natural sciences, the regulation of bodies in the interests of industrial efficiency and finally the rise of a money system whereby the value of all actions could be rationally calculated. That is, the development of an ethic of world mastery is simultaneously a historical process of rationalization and regulation in which the ethic of control involved the control of the flesh (the human passions), the regulation of the mind (by the development of formal systems of education), and finally the taming and training of the outer world of nature through an act of cultural colonization.[5] Just as there are three institutional orders, so there are three great problems confronting such arrangements. The regulation of the flesh is threatened by the ever-present problem of sexuality, the organization of the mind is constantly challenged by the presence of madness, and the colonization of reality (including other societies) is threatened by the ever-present problems of political resistance. The rationalism which Merton identified in the scientific experiment can be regarded as simply one instance of a broader and more comprehensive network of human practices which have the aim of world regulation.

As dimensions of an ethic of world mastery, the relationship between religion and medicine is both ancient and intimate. We may indicate this connection by taking note of the verbs to save the soul and to salve the body; the root notion of salvation was indeed *salus* indicating the most general condition of well-being. The connections between religion, medicine and control have deep roots within the classical Greek medical system. The notion of diet can be seen as a model for the government of the body and the government of society (Turner 1982). This perception of diet as both a model and metaphor of orderly social relations had its intellectual origins in the work of Michel Foucault on the character of Greek medical practice in relation to the emergence of Western notions of the self (Foucault 1987; 1988). Thus Foucault, commenting upon the work of Hippocrates, noted that

> the author of the treatise on *Ancient Medicine*, far from considering regimen as an adjacent practice associated with the medical art – one of its applications or extensions – attributes the birth of medicine to a primordial and essential preoccupation with regimen. According to him, mankind set itself apart from animal life by means of (a) sort of dietary disjunction.
>
> (Foucault 1987:99)

Etymologically, the notion of regimen is closely associated with the antique notion of regime, as a government or ordering of human relationships. These etymological connections also provide, of course, part of the groundwork for the many metaphors of the state and body which exist in European history; to order the life of men through a medical regimen is to provide the basis therefore for stable government and orderly legal relationships. Within the paradigm outlined by Foucault, we can plausibly regard medical practice as also a juridical-political regulation of society.

CRITICISM OF THE MERTON THESIS

While the Merton thesis has been influential in approaches to the history of science over a considerable period of time (Barber 1952; Merton 1984), it has also been a focus of critical evaluation from a variety of perspectives (Abraham 1983; Becker 1984). The Merton thesis raises a number of difficulties in terms of a comparative sociology of science. For example, Merton has somewhat neglected the contribution of Catholicism to scientific change (Feldhay 1988) and, specifically, the sociology of religion has failed to consider the consequences of the counterreformation for the development of Western rationalism.[6] One exception to this argument may be the work of Talcott Parsons (1971), in his comparative studies of the historical formation of societal communities. From a comparative perspective there is the well-known question of the contribution of French Jansenism to the evolution of a rational perspective as a consequence of the 'hidden God' problem in Jansenist theology (Goldmann 1964). There is also a more extensive difficulty in the case of China. While Weber had argued that the religious traditions of China were inimicable to the development of rational science (Weber 1951), it is clearly the case that China had a long-held and distinguished place in the historical development of science, making extensive contributions to basic science, medicine and technology (Needham 1954; Unschuld 1985). Furthermore, while it is a controversial issue, Merton's thesis also runs into difficulty in the case of Islamic sciences (Turner 1987b).

Merton's original interpretation of Weber has also been drawn on to question as a result of the ongoing debate on, not only the character of Weber's sociology, but more specifically on the nature of the sociology of science. Writers such as F. H. Tenbruck (1974) have given greater emphasis to the irrational roots of the Protestant drive for science, whereas Merton saw an affinity between the rationality of the Protestant life-world and the fundamental requirements of experimental science. A powerful alternative to the Weber-Merton thesis was presented by L. S. Feuer in *The Scientific Intellectual* (1963). Feuer attempted to locate the origins of rational science in the psychology and emotional structure of humanity. On this basis he attempted

> to show that the scientific intellectual was born from the hedonistic-libertarian spirit which, spreading through Europe in the sixteenth and seventeenth centuries, directly nurtured the liberation of human curiosity. Not asceticism but satisfaction; not guilt but joy in the human status; not self-abnegation, but self-affirmation; not original sin, but original merit and worth; not gloom, but merriment; not contempt for one's body and one's senses, but delight in one's physical being.
>
> (Feuer 1963:7)

Feuer argued, for example, that Galileo's scientific observations and investigations represented a challenge to the Church's authoritarianism by bringing about a psychological revolution which generated a new respect for the body, the emotions and the sensual life. In general, Feuer argued that the individualism of the seventeenth-century scientific revolution was essentially hedonistic, not ascetic.[7]

Feuer (1979) went on not only to reassert his thesis with considerable historical evidence, but also to call into question the entire historical basis for Merton's thesis, especially in terms of the membership of the Royal Society.[8]

THE HUMAN FABRIC

While Feuer attempted to support his argument by a number of empirical cases, his account of Andreas Vesalius's *De Humani Corporis Fabrica* of 1543 is of particular interest as a perspective by which to pursue the Merton thesis. For Feuer, 'the Vesalian problem' is an important illustration of the hedonistic-libertarian perspective which, according to Feuer's thesis, transformed social attitudes towards the human body. Following Charles Singer and Henry Sigerist (1924), Feuer regarded Vesalius's *Fabrica* as a revolutionary change, not only in the history of medicine, but within the whole evolution of discourses on the human body and on the place of anatomy within scientific medicine. It has been commonly observed that Christian asceticism and its teaching on the body as flesh resulted in the suppression of anatomical dissections for purely scientific objectives.[9] Given the Catholic antipathy to dissection, Feuer noted that there was widespread public opposition to Michelangelo's dissection of cadavers and that Pope Leo X closed Rome's hospitals to Leonardo da Vinci, which provided the immediate context for Leonardo's departure from Rome in 1515.

Feuer correctly drew attention to the conventional character of dissection and anatomy teaching within the Greek legacy of Galen (129–199 AD). Galen's text *On the Conduct of Anatomy* had become the authoritative source of medical understanding of the structure and functions of the body until it was eventually replaced in the late-sixteenth century by the development of pathological anatomy based upon direct observation of cadavers. The scholastic nature of the Galenic legacy can be illustrated by the fact that Galen never conducted an anatomical dissection of a human corpse, basing his knowledge instead on the bodies of monkeys; there is also some evidence that Galen undertook anatomical dissections of marine animals. Furthermore, traditional illustrations of anatomy lessons showed the Chief Surgeon or Professor of Medicine reading from the Galenic text, while an assistant conducted dissections from a corpse. These illustrations typically underscore the scholarly and theoretical authority of the anatomical text over direct observation and experiment. It is against this background that Andreas Vesalius Bruxellensis can appear as a revolutionary figure, since the *Fabrica* was grounded in actual anatomical dissections which were illustrated in his text. There were of course important precursors of Vesalius, including Berengario de Capri (Bologna) whose *Commentaria Cum Additionieus Super Anatomian Mundini* showed that Berengario had been working with actual dissections which conformed to his slogan ('experience of my eyes is my guiding star'). The importance of experiment had also been supported by Estienne in *De Dissectione Partium Corporis Humani Libri Tres* of 1539. However, the significance of Vesalius was his overt willingness to question the Galenic legacy by publishing his experimental results and illustrating his findings with clear educational illustrations.

The creation of pathological anatomy as a fundamental aspect of the medical curriculum can be regarded as one dimension of a wider movement of empiricism in science; as a revolt against Galenic deductivism. This empiricism was cogently supported in

England in the work of Francis Bacon in *The Advancement of Learning* in 1605 and by Robert Boyle's experimentations. The empirical approach was also noticeable in the clinical methods of Thomas Sydenham and John Locke (Cranston 1975; Payne 1900). However, one should not exaggerate the extent of this empiricism in actual practice. While Boyle had declared that he received more pleasure from a skilful anatomy than from examining the famous clock at Strasbourg, both Sydenham and Locke were critical of the actual utility of systematic, comparative anatomy. Sydenham suspected that the use of microscopes was immoral and was incompatible with God's purpose. It was the role of the physician to study 'the outer husk of things', not their inner workings. In fact, they associated comparative anatomy with theoretical medicine and mere book learning, claiming that anatomy should be subordinated to clinical practice. While the clinical method was grounded in the knowledge of proximate causes, comparative anatomy was associated in their minds with the impractical and false quest for ultimate causes (King 1970). The extent and nature of Locke's empiricism is therefore open to dispute (Soles 1985). However, in English medical circles, the importance of a sound knowledge of anatomy was widely accepted by the 1660s, especially under the influence of William Harvey's *De motu cordis* (1628) and *De circulatione* (1649) (Bylebyl 1979).

REMBRANDT'S ANATOMY LESSON

Feuer's thesis raises some important problems for the Mertonian perspective. If the human body is flesh (and the seat of evil passions), can the anatomical enquiry into dead flesh reveal the spiritual working of God in the universe? Did the revolution brought about by Vesalius represent a hedonistic challenge to the authoritarian structure of Christian theology, especially in its Catholic framework? Since both Catholic and Protestant theology regarded the body as the vehicle of human sin and failing, is it possible to differentiate between Protestant and Catholic conditions for the emergence of experimentally based comparative anatomy? In this section I propose to explore some of these issues by initially outlining the sociological interest of Rembrandt's famous painting of the anatomy lesson of Dr Nicolaas Tulp in the Waaggebouw in 1632 (Turner 1986).

The work of Rembrandt has had a fascination for sociologists and historians of art alike (Simmel 1985). Rembrandt's art

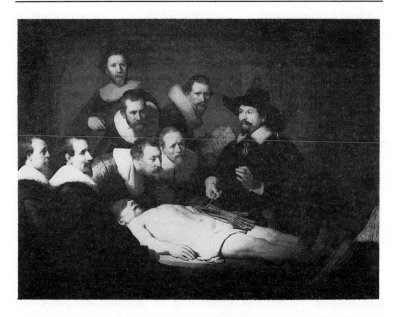

Rembrandt, *The Anatomy Lesson of Dr Tulp* (1632). © Mauritshuis, The Hague

represents a fusion of the nationalist-bourgeois sentiments of the Dutch War of Liberation, the expansion of Dutch commercial power, the realism of Caravaggio and a powerfully humanistic version of Christianity.[10] While Rembrandt painted two collective portraits of an anatomy lesson, the most famous is that depicting the dissection performed by Nicolaas Tulp in 1632 which is now housed in the Mauritshuis in Den Haag.

The painting has a number of interesting aspects from our point of view and my commentary largely follows the iconographic study by W. S. Heckscher (1958). The composition and style of the painting have striking features. For example, the contrast of light and darkness appears to follow closely the technique of Caravaggio, but symbolically it also represents the intense light falling upon the faces of the assembled surgeons, while the face of the corpse (one Aris Kint) is cast in shadow, indicating death. There are a number of conventional features to the painting which indicate its lineage with previous representations of anatomy lessons. For example, Dr Tulp, seated within the chair of authority, conducts his anatomy lesson *ex cathedra*. Behind his head there is set into the wall of the anatomy theatre the shell symbol of Christianity, so that in this painting we find an integration of the religious symbolism of authority and knowledge with the rational science of comparative anatomy. While most conventional representations of anatomy had shown the dissection of the abdomen, in this painting we find Tulp dissecting the hand and arm of the corpse. It may be that this indicated the descent of the anatomy lesson from the skills of Vesalius who was famous for his dissection of the hands, but interestingly in this representation of the anatomy lesson the arm of the cadaver is in fact both out of proportion and out of position. This provides an important clue to the real significance of the painting which is in many respects conventional, in line with Galenic rather than Vesalian principles. However, Tulp, like many contemporary anatomists, was significantly influenced by Andreas Laurentius (1558–1609) for whom the anatomy lesson had a moral purpose, namely as an instruction in the maxim 'know thyself'. The hand was regarded as a symbol of God's wisdom, hence to know the hand was to know God (Schupbach 1982). Commenting on the symbolic significance of the structure of this painting, Francis Barker (1984) has noticed that the eyes of the assembled surgeons do not in fact gaze upon the body of the dead man, but are directed instead to an anatomical text which is lodged in the bottom, righthand corner of the painting. The artistically dramatic presence of the body at the base of the pyramid of the surgeon and observers is, in a sense, denied by the gaze of the participants which is concentrated instead on a traditional text, but Barker's interpretation can be criticized on a variety of grounds (Christie 1986). For example, the picture which now stands in the Mauritshuis was repainted on several occasions. The original painting (as revealed by X-ray) shows an inner and outer triangle of surgeon-spectators. The inner triangle 'attend eagerly to various aspects of the demonstration, and therefore remain mentally within the picture' (Schupbach 1982:1). They are part of the ethical discourse of the painting. In short, the anatomy lesson is not a lesson about anatomy, but a discourse about medical and bourgeois authority over the disruptive forces of human disease, frailty and error. The anatomy lesson is a moral tale. Indeed, there is an important historical continuity between the medieval idea of the skeleton as the symbol of human frailty and the anatomical theatres which began to appear in the scientific institutions of northern Italy at the end of the sixteenth century. These have been appropriately referred to as 'amphitheatres d'anatomie moralisée' (Scheurleer and Meyjes 1975:221). Put within this broader historical context, we can now see the anatomy lesson is an important component of the baroque rhetoric of death (Chastel 1954),

THE CRIMINAL BODY

We have seen that while Merton in part derives the scientific ethic from the inner-worldly asceticism of Puritan Christianity, Feuer argues that the revolution brought about by surgeons like Vesalius was based on an ethic of hedonistic utilitarianism, which involved a celebration rather than a denial of our sensual existence. I have argued that we should regard the Merton thesis as one component within a more general notion of the ethic of world mastery, and my argument is that the anatomy lesson in particular can be seen within the Merton thesis as an illustration of Puritanical mastery over the world. My argument is therefore a modest defence of the Merton thesis against the criticism of Feuer, and I conclude by suggesting that Feuer has neglected one crucial feature in the development of comparative anatomy, namely that the bodies which were dissected traditionally were criminals who, having been condemned by some civic authority, were transferred from the scaffold to the anatomical theatre. The court room, the scaffold and the anatomical theatre were merely different locations of a single discourse of punishment and therefore scenes within a unified drama of destruction. The anatomy lesson was a juridical, moral lesson and only secondarily a scientific enquiry. In the seventeenth century, as in previous epochs, the juridical and the scientific were merged within a general cultural framework which was still primarily, indeed essentially, religious. It was not until the end of this period of the Golden Age that there was some degree of secularization: 'traditional theology was increasingly displaced from the early eighteenth century onward, by modern natural science as the chief soteriological bridge toward the understanding of the transcendental – and still Christian-Source of meaning' (Heyd 1988:176).

Historically, there have been basically two forms of anatomical dissection. There were private anatomical inspections of the dead within the aristocracy, when there was a suspicion of unnatural death, especially associated with poisonings. Second, there were the irregular anatomies of criminals throughout medieval times;

these dissections were not, however, directed by any significant or systematic medical interest in scientific experimentation. The development of anatomy is thus associated with the growth of professional medicine, with scientific experimentation, and with the practice of military surgeons such as Ambrose Paré (1510–90). While it was often difficult to secure a regular supply of cadavers by legitimate means, there appears to be a tradition whereby cadavers for anatomical experimentation and investigation were typically those of criminals. The public dissection (as in the case of Rembrandt's painting of Dr Tulp's lesson) was characteristically performed on a criminal body and therefore we can regard the annual public anatomy lesson as an aspect of a broader juridical punishment of the criminal. The punishment which had taken place on the gallows was continued in the form of an examination of the body of the criminal who was, as it were, slowly destroyed in the process of deconstruction as one feature of the legal system of institutionalized revenge. We can therefore interpret the anatomical investigation as part of the historical development of discipline within the framework of Foucault's investigations into the general evolution of disciplinary institutions (Foucault 1977).

In pre-modern Western political thought, the body of the king was the actual site of political power. The maintenance and reproduction of power was invested in a series of rituals which surrounded the public appearances of the king (especially coronations, royal entries and funerals). These public ceremonials were essentially liturgical and they came eventually to focus on the juridical theory of 'the King's two bodies' (Kantorowicz 1957). The king was regarded as having two bodies, one which was mortal and destructable, and a mystical body, which embraced and expressed the abstract sovereignty of the realm. Thus, at the burial of the king, the dead, decaying body was confined to the ground, while the lifelike effigy of the king was carried above, celebrating the unchallenged continuity of collective power as represented by the mythical political body of the king (Giesey 1987).

Because the king's body was both symbolically and factually so intimately bound into the institutions and symbols of state authority, any attack on the body of the king was necessarily an attack upon the body of society. The practices and symbolism of the scaffold therefore perfectly expressed this idea that regicide was the most serious attack upon the very foundations of orderly, civilized life. The hanging, drawing and quartering of such regicides symbolized their total exclusion from human society, since, given the doctrine of the resurrection of the body, the criminal was killed in this world in such a way as to prevent his restoration at the Second Coming. They died therefore a double death, being excluded from all possible forms of human society.

The anatomy lesson as an extension of the public ceremonial of the scaffold thus represents a scientific destruction of the body of the criminal in the interests of the body of the king. The anatomy theatre is in this respect the counter-image of the coronation ceremony. It represents the ethic of world mastery translated and transferred into the political arena, where the unmasterly wills of recalcitrant humans are finally subordinated and disintegrated by the physical intervention of instrumental reason through the hands of the surgeon, operating simultaneously as the guardian of knowledge and power. Although the doctrine of the resurrection of the body, the use of the body as a metaphor of the political, the idea of the medical regime as a model of social organization, and the legitimization of social differentiation by reference to the differentiation of hand and mind preceded the Reformation, having deep and important roots within Catholic culture, the Puritan Revolution gave these metaphors and practices a new direction; for example, they gave a greater emphasis to the notion of individual responsibility. The Puritan Revolution involved the idea of Man as a totally fallen creature, whose sins could not be resolved one by one (for example, through the confessional) but only by a total salvational act. The mastery of the flesh therefore required a massive, detailed and more subtle panoply of surveillance and regulation, of which the scientific probing of the body and the mind were crucial features of the Puritan Revolution. In this respect, the anatomy lesson can be seen as further vindication of Merton's original hypotheses.

CONCLUSION

It is often tempting to write about Weber's sociology as a form of simplistic functionalism; in the case of this analysis, it may appear that the Protestant ethic of world mastery functioned to promote the conditions favourable to the advance of science. However, it is clear that Weber's sociology cannot be simply reduced to some form of idealistic functionalism. In particular, Weber saw society as a fragmented array of loosely-knit and independently developing sectors in continuous competition and even conflict with one another.

(Kalberg 1987: 152)

Weber's sociology of meaningful action takes into account the importance of unintended consequences, the impact of contingent historical circumstances, the complexity of social change and the centrality of interpretation in the historical sciences. In a similar fashion, we should see the rise of anatomy as an aspect of the medical curriculum as the outcome of diverse historical circumstances including not only value complexes, but developments in military technology, institutional reorganization of the medical profession and competition between different schools of medical practice. Rembrandt's painting of Tulp's lesson tells us as much about municipal pride as it does about scientific advances in anatomical dissections.

Merton's historical sociology of scientific change can, in a similar fashion, be treated as a particularly valid illustration of theoretically sophisticated, middle-range empirical investigation of the general conditions for the emergence and support of instrumental rationality. However, one criticism of Merton's thesis which emerges from this note is that, at least in the early decades of the seventeenth century, there was relatively little differentiation between science, religion and law; we may, as a result, regard the anatomical dissection of cadavers as a feature of the juridical management of deviance.

NOTES

1 This chapter was originally given as a seminar paper to the Department of Anthropology, University of Adelaide. I would like to thank Professor Bruce Kapferer for his comments. Professor R.K. Merton kindly made available his Sarton centennial lecture from the University of Ghent. Finally, Dr Jan Rupp, University of Utrecht, has provided invaluable bibliographical sources. The errors are mine alone.

2 In many respects, the sociology of salvational drives and institutions across different religious cultures and traditions is still an underdeveloped aspect of the sociology of religion. Some comparative components of the salvational structures of moral discourse have been analysed in the historical sociology of consciousness (Hepworth and Turner 1982; Hodgson 1974; Huff 1981).

3 In traditional religious and medical systems, the regulation of the flesh was often achieved through diet. Foucault's (1987) study of medical

practice and dietary management is of particular interest in this context. However, Foucault's gender blindness had important implications for his approach to these questions.

4 The complicated debate which surrounds the Weber thesis itself is not an essential feature of this commentary on Merton's sociology of science. I shall have to assume the general validity of Weber's own argument, thereby ignoring contemporary criticisms (MacKinnon 1988).

5 The intellectual connection between Adorno and Horkheimer's critique of capitalist regulation of the life-world and Weber's story of the Protestant Ethic has been generally neglected. Within a broader spectrum, the philosophical connections between the radical and the conservative critique of instrumental rationality have been equally ignored or suppressed. The link between Weber and Adorno may well be the conservative and romantic thinker Ludwig Klages, who thought that the cosmic rhythm of the human body and nature had been destroyed by the instrumental rationalism of capitalism (Klages 1963).

6 One exception in the sociology of religion to this general rule is the work of Swanson (1967).

7 It is important to compare this analysis of science and individualism with other approaches which consider the relationship between economics and individualism (Abercrombie *et al.* 1986; Chenu 1969; MacFarlane 1978).

8 It is interesting to note that, in his authoritative study of Robert K. Merton, Sztompka (1986) makes no reference to Feuer's critique.

9 Some important aspects of the history of anatomy are discussed in Castiglione (1941), Edelstein (1935), Farrington (1932) and Wolf-Heidegger and Cetto (1967). Various aspects of the contradictions between the body, health and spirituality are revealed in recent studies in the history of anorexia nervosa (Bynum 1986; Bell 1985).

10 The symbolism and culture of the urban patriciate of the United Provinces in the seventeenth century are outlined in detail in Schama's *The Embarrassment of Riches* (1987). An overview of the Dutch Revolt is presented in Parker (1985). The general social context of the universities and intellectuals is exhaustively documented by Frijhoff (1982; 1983). The main features of scientific development in the Netherlands are analysed in Hackmann (1975).

REFERENCES

Abercrombie, N., S. Hill and B.S. Turner (1986) *Sovereign Individuals of Capitalism*, London: Allen & Unwin.

Abraham, G.A. (1983) 'Misunderstanding the Merton thesis, a boundary dispute between history and sociology', *ISIS* vol. 74:368–787.

Adorno, T. and M. Horkheimer (1979) *Dialectic of Enlightenment*, London: NLB, Verso edition.

Barber, B. (1952) *Science and the Social Order*, Glencoe, Illinois: The Free Press.

Barker, F. (1984) *The Tremulous Private Body, Essays on Subjection*, London and New York: Methuen.

Becker, G. (1984) 'Pietism and Science: a critique of Robert K. Merton's hypothesis', *American Journal of Sociology* vol. 89(5):1065–90.

Bell, R.M. (1985) *Holy Anorexia*, Chicago: Chicago University Press.

Bylebyl, J.J. (ed.) (1979) *William Harvey and His Age: the Professional and Social Context of the Discovery of the Circulation*, Baltimore and London: Johns Hopkins University Press.

Bynum, C.W. (1986) *Holy Feast and Holy Fast: the Religious Significance of Food to Medieval Women*, Berkeley: University of California Press.

Castiglione, A. (1941) 'The origin and development of the anatomical theater to the end of the Renaissance', *Ciba Symposia* vol. 3:826–44.

Chastel, A. (1954) 'Le Baroque et la mort', *Retorica e Barocco*, Attidel III, Congresso Internazionale di Studi Umanistici: 33–46.

Chenu, M-D. (1969) *L'éveil de la conscience dans la civilisation médiévale*, Montreal: Institut d'Etudes Médiévales.

Christie, J.R.R. (1986) 'Bad news for the body', *Art History* vol. IX: 263–71.

Cranston, M.W. (1975) *John Locke: A Biography*, New York: Macmillan.

Edelstein, L. (1935) 'The development of Greek anatomy', *Bulletin of the Institute of the History of Medicine* vol. 3: 235–48.

Farrington, B. (1932) 'The Preface of Andreas Vesalius to *De Fabrica corporis Humani* 1543', *Proceedings of the Royal Society of Medicine* 25:1357–66.

Feldhay, R. (1988) 'Catholicism and the emergence of Galilean science: a conflict between religion and science?', *Knowledge and Society: Studies in the Sociology of Culture Past and Present* vol. 7:139–63.

Feuer, L.S. (1963) *The Scientific Intellectual: The Psychological and Sociological Origins of Modern Science*, New York: Basic Books.

Feuer, L.S. (1979) 'Science and the ethic of protestant asceticism: a reply to Robert K. Merton', *Research in Sociology of Knowledge: Sciences and Art*, a research annual vol. 2:1–23.

Foucault, M. (1977) *Discipline and Punish: Birth of the Prison*, London: Tavistock.

Foucault, M. (1987) *The History of Sexuality; Volume 2: The Use of Pleasure*, Harmondsworth: Penguin Books.

Foucault, M. (1988) 'Technologies of the self', pp. 16–49 in L.H. Martin, H. Gutman and P.H. Hutton (eds), *Technologies of the Self: A Seminar with Michel Foucault*, London: Tavistock.

Frijhoff, W. Th. M. (1982) 'De arbeidsmarkt voor academici tijdens de Republiek', *Spiegel historiael* vol. 17(10):501–10.

Frijhoff, W. Th. M. (1983) 'Wetenschap, beroep en status ten tijde van de Republiek: de intellectueel', *Tijdschrift voor de geschiedenis der geneeskunde, natuurwetenschappen, wiskunde en techniek* vol. 6:18–30.

Gehlen, A. (1988) *Man, his Nature and Place in the World*, New York: Columbia University Press.

Giesey, R.E. (1987) 'The King imagined', pp. 41–60 in K.M. Baker (ed.) *The French Revolution and the Creation of Modern Political Culture, vol. 1. The Political Culture of the Old Regime*, Oxford: Pergamon Press.

Goldmann, L. (1964) *The Hidden God*, London: Routledge & Kegan Paul.

Hackmann, W.D. (1975) 'The growth of science in the Netherlands in the seventeenth and early eighteenth centuries', pp. 89–109 in M. Crosland (ed.) *The Emergence of Science in Western Europe*, London: Macmillan.

Heckscher, W.S. (1958) *Rembrandt's Anatomy of Dr. Nicolaas Tulp: An Iconological Study*, New York: New York University Press.

Hepworth, M. and B.S. Turner (1982) *Confession: Studies in Deviance and Religion*, London: Routledge & Kegan Paul.

Heyd, M. (1988) 'The emergence of modern science as an autonomous world of knowledge in the Protestant tradition of the seventeenth century', *Knowledge and Society: Studies in the Sociology of Culture Past and Present* vol. 7:165–79.

Hodgson, M.G.S. (1974) *The Venture of Islam*, Chicago: Chicago University Press; 3 volumes.

Huff, T.E. (ed.) (1981) *On the Roads to Modernity Conscience: Science and Civilizations, Selected writings by Benjamin Nelson*, Totowa, NJ: Rowman and Littlefield.

Kalberg, S. (1987) 'The origin and expansion of Kulturpessimismus: the relationship between public and private spaces in early twentieth-century Germany', *Sociological Theory* vol. 5(2):150–64.

Kantorowicz, E.H. (1957) *The King's Two Bodies*, Princeton, New Jersey: Princeton University Press.

King, L.S. (1970) 'Empiricism and rationalism in the works of Thomas Sydenham', *Bulletin of the History of Medicine* vol. 44(1):1–11.

Klages, L. (1963) *Vom kosmogonischen Eros*, Bonn: Bouvier.

MacFarlane, A. (1978) *The Origins of English Individualism*, Oxford: Blackwell.

MacKinnon, M.H. (1988) 'Part I: Calvinism and the infallible assurance of grace', *British Journal of Sociology* vol. 34(2):143–77.

Merton, R.K. (1938) 'Science, technology and society in seventeenth century England', *Osiris: Studies on the History and Philosophy of Science, and on the History of Learning and Culture* vol. IV(2):360–632.

Merton, R.K. (1970) *Science, Technology and Society in Seventeenth-Century England*, New York: Howard Fertig.

Merton, R.K. (1984) 'George Sarton: episodic recollections by an unruly apprentice', mimeographed.

Needham, J. (1954) *Science and Civilization in China*, Cambridge: Cambridge University Press; 7 volumes.

Parker, G. (1985) *The Dutch Revolt*, London: Allen Lane.

Parsons, T. (1971) *The System of Modern Societies*, Englewood Cliffs: Prentice-Hall.

Payne, J.F. (1900) *Thomas Sydenham*, London: Unwin.

Poggi, G. (1983) *Calvinism and the Capitalist Spirit: Max Weber's Protestant Ethic*, London: Macmillan.

Schama, S. (1987) *The Embarrassment of Riches: an Interpretation of Dutch Culture in the Golden Age*, New York: Alfred A. Knopf.

Scheurleer, Th. H.L. and G.H.M.P. Meyjes (eds) (1975) *Leiden University in the Seventeenth Century: An Exchange of Learning*, Leiden: E.J. Brill.

Schupbach, W. (1982) *The Paradox of Rembrandt's 'Anatomy of Dr Tulp'*, London: Wellcome Institute for the History of Medicine.

Simmel, G. (1985) *Rembrandt, ein Kunstphilosophischer Versuch*, Munchen: Matthes and Seitz.

Singer, C. and H.E. Sigerist (1924) (eds) *Essays on the History of Medicine*, London: Routledge & Kegan Paul.

Soles, D.E. (1985) 'Locke's empiricism and the postulation of unobservables', *Journal of the History of Philosophy* vol. 23:339–69.

Stauth, G. and B.S. Turner (1988) *Nietzsche's Dance, Resentment, Reciprocity and Resistance in Social Life*, Oxford: Basil Blackwell.

Swanson, G.E. (1967) *Religion and Regime*, Ann Arbor, MI: Michigan University Press.

Sztompka, P. (1986) *Robert K. Merton: An Intellectual Profile*, London: Macmillan.

Tenbruck, F.H. (1974) 'Max Weber and the sociology of science: a case reopened', *Zeitschrift für Soziologie* vol. 3:312–23.

Turner, B.S. (1982) 'The government of the body: medical regimens and the rationalization of diet', *British Journal of Sociology* vol. 33(2):252–69.

Turner, B.S. (1984) *The Body and Society: Explorations in Social Theory*, Oxford: Basil Blackwell.

Turner, B.S. (1986) 'Foucault and the crisis of modernity', *Theory, Culture & Society* vol. 3(3):179–82.

Turner, B.S. (1987a) *Medical Power and Social Knowledge*, London: Sage.

Turner, B.S. (1987b) 'State, science and economy in traditional societies: some problems in Weberian sociology of science', *British Journal of Sociology* vol. 38(1):1–23.

Unschuld, P.U. (1985) *Medicine in China: a History of Ideas*, Berkeley: University of California Press.

Weber, M. (1951) *The Religion of China*, New York: Macmillan.

Weber, M. (1966) *The Sociology of Religion*, London: Methuen.

Wolf-Heidegger, L. and A.M. Cetto (1967) *Die Anatomische Sektion in Bildlicher Darstellung*, Basel: S. Kargen.

Zola, L.K. (1972) 'Medicine as an institution of social control: the medicalizing of society', *The Sociological Review* vol. 20(4):487–504.

Chapter 8

The talking disease
Hilda Bruch and anorexia nervosa

INTRODUCTION

In this discussion of anorexia nervosa, it is assumed that the sociology of health and illness must adopt a three-level model of explanation. We have to understand the phenomenology of illness within the world of the patient. The problem with 'positivistic' approaches (based on a medical model) is that they examine the condition independently of its constitution by a knowledgeable agent. Anorexia is not just a question of having anorexia; it is fundamentally about being anorexic. Secondly, there is the sociology of illness behaviour which examines the sick role. Finally, there is a political economy of health which is broadly concerned with the distribution of resources (including health) in a society. It can also be argued that a 'condition' as complex as anorexia requires a multidisciplinary perspective to understand the various levels in which this disorder can be conceptualized in terms of its various social and cultural meanings. Existing approaches to anorexia are typically underdeveloped because they adopt uncritically a unidimensional view of the aetiology of anorexic behaviour. Hilda Bruch's clinical work is important in this context as a multidisciplinary interpretative approach. As a supplementary theoretical strategy, this discussion extends the therapeutic insights of Bruch by considering anorexia in terms of a phenomenology of the mouth as a 'talking disease' about familial interaction in societies which place an emphasis on individual competition. Being sick involves a special type of linguistic membership through socialization in a sick role. The understanding of the meaning of sickness in medical practice can be seen consequently as a version of the methodology of *verstehende soziologie*.

In this discussion my aim is, through a review of various contemporary approaches to 'eating disorders', to suggest that becoming sick is like becoming a member of a social (and therefore linguistic) community. Becoming sick requires the patient to learn how to perform according to certain norms of appropriate behaviour; there are appropriate and inappropriate rules of conduct which are expected of a person who has been correctly identified as 'sick'. To understand a sickness is parallel to the anthropological task of understanding an alien culture because it requires an interpretation of many competing signs which may stand for, but also disguise, a range of possible medical conditions. There may also be, so to speak, ethnomethodological uncertainty as to the significance and seriousness of the 'complaint' for both victim and observer. In short, the language by which 'victims' describe their complaints is an essential component of interpretation for both sociologist and clinician. The goal of medical sociology can be seen therefore as one specific instance of the general aim of sociology itself which is to understand the meaning of social action in its sociocultural setting (Turner 1987a). In terms of the sociology of symbolic communication, this commentary will attempt *inter alia* to establish certain relationships between the meaning of eating and talking. This approach to illness was first systematically outlined in Maurice Merleau-Ponty's phenomenology of perception (1962) in which apprehending reality was first grounded in the presupposition of an embodied social agent.

Throughout its nosological history a disease or sickness will, often in a paradoxical and ironic fashion, summarize and articulate the personal problems associated with contradictions and strains in culture and social structure. This notion formed the basic insight in Talcott Parsons's attempt to understand the relationship between the sick role and strains in the social system (Parsons 1951; Gerhardt 1989; Holton and Turner 1986). Let us consider the following examples. In medieval times leprosy and epilepsy (the so-called sacred diseases) were used simultaneously to give expression to the Church's horror at the corruption of the human body and to illustrate the fact that God educated the soul through the suffering of the body (Turner 1984). In the seventeenth century, Robert Burton in *The Anatomy of Melancholy* attempted to articulate the Baroque features of human sadness in his analysis of melancholic nostalgia. The confused discourse of *The Anatomy* reflected and mimicked the disorder of the condition of melancholy (Lyons 1971;

Fox 1976). According to Michel Foucault in *Madness and Civilization* (1967), Descartes sought to exclude madness from his rationally mechanical universe in order to preclude instabilities from a world of positivistic coherence, but Descartes also saw his task as contributing to the health of humanity. Towards the end of the nineteenth century, hysteria and agoraphobia were constituted as metaphors of the peculiar problems of the urban, middle-class family in which disruptive sexuality constantly threatened the formal stability of bourgeois life (de Swaan 1981).

Joan Brumberg in her historically sensitive and scholarly study of *Fasting Girls* (Brumberg 1988) rightly notes that anorexia nervosa was the disease of the 1970s which, at least in the public imagination, was expressive of concerns with consumption, personal display, feminist politics, the fashion for dieting and exercise, the individualistic competitiveness of advanced capitalist societies and finally one aspect of the larger debate surrounding the notion of the narcissistic self. I shall return later to the aetiological complexity of anorexia. The AIDS epidemic expresses both the global spread of disease and the problems of congested spaces and uncontrolled, anonymous intimacies, giving rise to greater demands for the surveillance and control of secret sexuality (Sontag 1989). In short, disease categories which mark out the division between the amoral causality of nature and the social world of moral actions have been important features of political discourse throughout human history, because they perfectly express the problem of the transgression of boundaries; they articulate the Outside and the Other (Boyne 1990).

From a scientific point of view the 'epidemic' of anorexia in the 1970s raised important questions about the culture-boundedness nature of disease categories, since anorexia (when narrowly and clinically defined) appeared to be specific to affluent, middle-class, urban, Western cultures; approximately 95 per cent of anorexics are young, female, white and from predominantly middle- or upper-class families (Prince 1985; Varma 1979). If we treat illness as a symbol not only of disorders in the patient but as metaphors of social arrangements which have gone awry, then we need methods of reading discourses of disease (Sontag 1978; Turner 1982). For example, it is possible to think metaphorically of some diseases as diseases of distinction. In the nineteenth century, tuberculosis was often associated in literary circles with intellectual and artistic activity, while the notion of the hyperactive child in the twentieth century has been connected with the idea of a surplus of intelligence.

In American popular literature, anorexia was made 'fashionable' by the tragic death in 1983 of Karen Carpenter and by the confessions of Jane Fonda to a period of bulimia (literally 'ox hunger'). Of course, medical historians are also conscious of the fact that it is possible to argue that certain exemplary medieval saints (such as Catherine of Siena) exhibited anorexic behaviour (Bell 1985; Brown 1981; Bynum 1984 and 1985). Fasting as part of an ascetic life-style has been of course almost universal in religious communities (for example in Theravada Buddhism). The Egyptian desert fathers also adopted many of the ideas and practices of ancient Greek medicine in their development of a dietary discipline; these medico-religious systems were eventually further elaborated in early classical times (Feher *et al.* 1989). However, we should be suitably cautious about making such large, transhistorical and crosscultural claims about the historical continuity of a 'disease' such as anorexia, precisely because of the cultural specificity of the symptomatology of human disorders. Indeed, this problem of comparison and continuity raises epistemological difficulties about the stability of the relationship between signified and signifiers which lies well beyond the scope of this modest note.

DEFINING THE ANOREXIC STATE

What then is anorexia nervosa? Brumberg provides us with a particularly useful account of the historical concretion of the disease category. The emergence of anorexia nervosa as a modern disease can be dated to a diagnostic description by Sir William Withey Gull, physician to the family of Queen Victoria, who in 1873 designated an emaciated patient as 'an extreme instance of what I have proposed to call apepsia hysterica or anorexia nervosa'. In France, Charles Lasegue in 1873 produced a description of *l'anorexia hysterique* which provided the first real insight into the familial origins of the disease, which was regarded as being triggered by an emotional crisis associated with the blocking of expectations. In particular, he associated anorexia with the frustration of unfulfilled sexual expectation in courtship and early married life. Both Jean-Martin Charcot and Sigmund Freud connected *l'anorexia hysterique* with sexual frustration and with the intensity of middle-class management of maturing females, and proposed the removal of the anorexic daughter from the family household as the principal therapeutic intervention. In the nineteenth century, prior to its precise medical definition, anorexia was also either confused or associated with dyspepsia and with chlorosis (a type of anaemia named after its greenish tinge which supposedly characterized the skin of the patient, and which was also occasionally referred to as Virgin's Disease: Loudon 1980; McFarland 1975; Siddall 1982). These disorders were occasionally regarded as forms of cultivated fragility or emotional daintiness. These female digestive problems were sometimes 'cured' by some magical tonic or fashionable medicine such as Dr Williams' Pink Pills for Pale People (Brumberg 1988:173). The maturing but unstable sexual appetite of the female had to be monitored and regulated by an appropriate diet: for example, it was well known that excessive meat-eating produced nymphomania. Michel Foucault has noted that, within a broader historical context, the 'association between the ethics of sex and the ethics of the table was a constant factor in ancient culture' (Foucault 1987:50). In the anthropological literature female abstinence is associated with liminal states, that is with boundary problems between giving and taking, eating and dieting, pregnancy and menstruation, eating and talking (Bynum 1988; Pina-Cabral 1986).

In the period 1900–1940, the therapeutic management of anorexics was dominated either by the biological or by the psychoanalytic model of disease. These attempts to organize anorexia as a distinctive medical condition by providing it with a scientific discourse and an appropriate therapeutics were also an important indication of the professional struggle between various medical and quasi-medical occupations for the dominance of a middle-class clientele. For example, the recognition of the significance of hormones and their clinical utility was an important step in the development of the medical treatment of the anorexic. In particular, anorexia was connected with both pituitary and thyroid dysfunctions. The results of organotherapy were, however, less than convincing. For instance, it proved difficult to achieve a successful clinical differentiation of Simmonds' disease and anorexia (Escamilla 1944). In 1942 a summary of the literature on Simmonds' disease showed that of approximately 600 reported cases only 100 were established through pathological evidence. Although many cases of Simmonds' disease were in fact misdiagnosed cases of anorexia, many emaciated women were treated with pituitary extracts, despite the fact that they showed no particular glandular insufficiency. It can be argued that any improvement in the patient's condition as a result of endocrine therapy was in fact brought about by a placebo effect which was associated with the hysterical nature of the anorexic responding temporarily to medical intervention.

The treatment of anorexia was transformed by three factors: the therapeutic failure of endocrinology, the growing influence of the Freudian psychoanalytic movement and finally the medical recognition of the importance of emotions in the origin and treatment of diseases. There were important developments in psychophysiological medicine in the 1930s which made general medical practice more open to the notion that there were important interactions between mind, body and social context (Alexander and Selesnick 1966). Growing recognition of the significance of psychosomatic medicine provided the basis for a more integrated and eclectic therapy for anorexia. In addition, by the 1970s the nosological map of the clinical signs and symptoms for a scientific diagnosis of anorexia had been more firmly established by the following criteria: (1) age of onset prior to twenty-five years; (2) with at least 25 per cent loss of original body weight; (3) the existence of a distorted attitude towards food and eating; (4) no known prior medical condition which could account for the presence of anorexia; (5) no other known primary affective psychiatric disorders; and (6) at least two of the following – amenorrhoea, lanugo, bradycardia, overactivity, bulimia and vomiting (Feighner 1972; Palmer 1980). It is now also common to identify both bulimia and anorexia as versions of a broader 'dietary chaos syndrome' (Palmer 1979). However, the successful treatment of anorexia remains an elusive and difficult therapeutic goal. It is for her contribution to therapy that the practice of Hilda Bruch (1904–1984) is universally recognized. The essence of her diagnostic approach was her painstaking attention to how her patients described their own world-view of the anorexic condition. What has not been adequately recognized, however, is the proximity between Bruch's clinical interpretations of the meaning of anorexia and the classical tradition of sociology which is grounded in the quest for interpretative understanding of action. In short, we have neglected the proximity between structuration theory and the clinical interpretation of being sick as social action, and thus between clinical and sociological interpretation.

THE GOLDEN CAGE OF THE ANOREXIC

Dr Bruch received her MD from the University of Freiburg in 1929, undertook physiological research training in Kiel and paediatric training in Leipzig, and became a refugee from Nazism, settling eventually in the United States, where she was Professor of Psychiatry at Baylor College of Medicine in Houston (Lidz 1985). She is widely recognized for her contribution to the diagnosis and treatment of eating disorders, publishing a number of influential studies such as *The Importance of Overweight* (1957); *Eating Disorders: Obesity, Anorexia and the Person Within* (1973); *Learning Psychotherapy* (1974); and *The Golden Cage, The Enigma of Anorexia Nervosa* (1977). Her *Conversations with Anorexics* (1988) were dictated during her final illness, and in some respects they represent an eloquent and moving summary of her life's work.

Bruch's argument (or rather the conclusion of her long and patient encounter with anorexics) is that neither behavioural-modification programmes nor family therapy will be therapeutically effective unless they seriously address the patient's deep, long-standing and persistent personality problems which are associated with a lack of personal autonomy, overconformity and an appalling lack of personal esteem. Existing therapies had failed because they did not address the question of the meaning of anorexia for the anorexic; that is, they had failed to grasp the phenomenology of what it was to be an anorexic rather than simply having anorexia. Two salient features emerge from Bruch's careful study of anorexics, namely the presence of an overpowering, dominant mother involved in an excessively regulated relationship with her daughter, where there is a contradictory stress on compliance, cleanliness and individual competitiveness, and secondly an inadequate familial preparation for adolescence, because the anorexic household offers few opportunities for individualization. The result is that these middle-class females find it difficult to leave the family and make the transition into a broader, more exciting but more demanding world. The aim of Bruch's therapy was to provide an insight into these constraining conditions which would ultimately provide the patient with personal empowerment.

The title of Bruch's earlier study (*The Golden Cage*) perfectly captures the paradoxical feelings of the anorexic daughter, namely that she is merely a sparrow in a golden cage, too plain to compete with her familial peers or with her mother, but also deprived of the freedom of doing what she really wants or of achieving what she really might be. We might reasonably argue that the aim of Bruch's therapy could be expressed in terms of the will to power to 'want to become those we are' (Nietzsche 1974, section 335). Self-imposed starvation provides a form of personal control which is expressive of a pseudo-power, but this artificial regime in fact prohibits the daughter's flight from the family cage. By suppressing menstruation through starvation and exercise, the daughter suppresses her own gender identity and sexual maturity. By adopting a permanently childlike body, voice and outlook, she precludes the possibility of successful adulthood. Anorexic behaviour is a personal response to the confusions and contradictions of female maturation which may be expressed in a series of dichotomies – personal autonomy/compliance, childhood security/mature independence, sexuality/neutrality.

Conversations with Anorexics (Bruch 1988) develops and deepens the insights of her previous work, by showing and exhibiting the complexities of the religio-moral world-vision of the anorexic, who typically regards food, eating and the body as morally or indeed spiritually degraded. Starvation and exercise represent, unconsciously or consciously, a negation of that corruption. The anorexic avoids the shameful world of eating, while simultaneously achieving personal power and a sense of moral superiority through the emaciated body. Their attempt at disembodiment through negation becomes the symbol of their moral empowerment. It is on this basis that we can connect the age-old practices of Western asceticism and saintship with the modern moral dilemma of Western affluence in a world of starving millions. The complexity of the contemporary symbolism of anorexia is that modern consumerism appropriates all forms of symbolism (including oppositional, anti-capitalist symbolism) to its own commercial purposes. Being hyper-slim, while in opposition to the signs of affluence, is also cool.

However, as Bruch successfully indicated to her patients, this 'solution' to the contradictions of the status of young women in modern societies represented merely a pseudo-solution based on a false appraisal of their place within the world. The goal of Bruch's therapy was to provide the patients with a genuine sense of their own autonomy and value, and therefore confidence in their ability eventually to leave the golden cage.

Modern philosophy has above all reinforced the idea that language is constitutive of the objective reality in which we are housed. It is the view of the relationship between words and things which provides one starting point for the thesis that diseases are socially constructed (Bury 1986). Furthermore, the philosophy of medicine has also been profoundly influenced by the recent revival of the study of the work of Ludwik Fleck (Cohen and Schnelle 1986). Medical categories for disease entities are themselves the product of 'thoughtstyles' (*Denkstil*) of scientific communities (*Denkgemein-schaft*). As a result, we have been sensitized to the metaphoricality of the social world, and this notion of metaphor is particularly important in the cultural understanding of disease. The importance of this approach owed a great deal to the work of Susan Sontag (1978, 1989) who proposed that we should regard illness as metaphor. In this respect, anorexia is a peculiarly articulate disease, despite the fact that in a sense the anorexic has no voice; that is, no articulate place in social space. Not eating expresses autonomy from parental demands, but it has the ultimate consequence of increasing dependence (on parents and professional help). As Brumberg correctly points out, nineteenth-century anorexia was associated with a far broader category of problems, namely the 'wasting diseases' of bourgeois women. In extreme cases, anorex-ics waste away as a consequence of regarding their own lives as a moral wasteland.

THE PHENOMENOLOGY OF ANOREXIA

Our understanding of anorexia can benefit from consideration of Merleau-Ponty's phenomenology of embodiment and the notion of the lived body (Merleau-Ponty 1962). For Merleau-Ponty, our bodies are never merely extensions in space, but a complex interaction between an environment (*Umwelt*), our sociocultural habitus (*Welt*) and our ongoing intentionality. Within this framework, 'eating disorders' are never merely 'events' within the *Umwelt* but belong to the intentionality of our own particular *Welt*. The intentional refusal to eat, the loss of appetite and the wilful vomit threaten to break social relations. There may be a parallel therefore between the loss of voice, the loss of individuality and the loss of weight. We might imaginatively at least draw a parallel between talking disorders and eating disorders. Merleau-Ponty, following L. Binswanger's *Über Psychotherapie* (1935), comments on the case of a young woman who, having been forbidden to see her lover, lost her voice and was unable to speak. He notes that the mouth is essential not only to sexual development but to social development through communication:

> In so far as the emotion elects to find its expression in loss of speech, this is because of all bodily functions speech is the most intimately linked with communal existence.
>
> (Merleau-Ponty 1962:160)

Loss of speech (or loss of appetite) is in this sense a profound act of communication; it signifies a departure from the sociolinguistic community of the healthy into the more privatized linguistic community of the anorexic (Caskey 1986). At the same time it is a 'moral' statement about the anxiety of maturation into selfhood (Eckerman 1987). Any therapeutic intervention will have to address these issues, namely the conditions under which an anorexic patient can be given a voice. This perspective provides the key to the failure of most conventional therapies which ultimately see the condition as a problem within the *Umwelt* of the patient and thereby deny her intentionality.

Bruch's clinical histories of anorexic patients thus raise a number of problems for both popular and professional approaches to anorexia. It is clearly the case that anorexia can be neither explained nor treated within the paradigm of biochemical, scientific medicine. For example, hormonal treatment is not effective over a long period. Anorexia is resistant to treatment, precisely because the anorexic perceives weight gain as a moral and personal failure. In addition, anorexia cannot be explained within an exclusively feminist paradigm, despite the fact that 95 per cent of anorexics are female. It is also the case that women 'are more concerned with their appearance than are men, since, traditionally, women have been judged on appearance more than men have' (Hayes and Ross 1987:124). The relationship between eating, health and appearance is particularly important for women, especially in a society which gives so much prominence to the representational self (Abercrombie *et al.* 1986). Although anorexia may well fit into a 'control paradox' as a consequence of the powerlessness of women in a patriarchal context (Lawrence 1979), it is too simplistic to argue that anorexia is an outcome of male-dominated notions of beauty and fashion. Like the corset, the anorexic body is a very complex statement of socio-moral worth (Davies 1982). The anorexic does not necessarily attempt to conform implicitly or explicitly to some male or patriarchal model of the beautifully slim, sexually attractive woman;

the conscious aim of much anorexic behaviour is to subordinate female sexuality, to deny the gender-specific characteristics of personality and to withdraw from any sexual contact with men (MacLeod 1981). In addition, from the point of view of family therapy, the principal dynamic in the social aetiology of this disease is the conflict between the daughter and the overprotective mother. The idea that anorexia is simply the product of a patriarchal-capitalist culture is too vague and imprecise to be of any particular interest. It is also not clear what the therapeutic implications of such a diagnosis would be. The anorexic diets not to become sexually available but to become socially unavailable and to communicate by not communicating. The discourse of their diet is one of social exclusion. However, we may to some extent regard anorexia as an ascetic and moral response to contemporary consumerism, partly because the autobiographical accounts of anorexia give special prominence to this association of obesity and moral worthlessness. Anorexia is the peculiar consequence of a culture fascinated by individual competition, dietary management, the narcissistic body and the presentational self, but Bruch's work also reminds us that anorexia is a specific personal response to the peculiarities of certain forms of moral management within the middle-class household. Bruch's diagnostic and therapeutic orientation to the anorexic via her personal biography, family history and class position had precisely the multidisciplinary approach necessary to obtain a comprehensive overview of the complaint, but she also combined this orientation with a clear understanding of the phenomenological complexity of being anorexic rather than simply having anorexia.

In this respect, we may talk about anorexia as an overdetermined disease, being the consequence of cultural, social, familial and maturational processes which create 'sick roles' for anorexic candidates. Anorexia should be approached conceptually and theoretically at three levels (Turner 1987b). At the phenomeno-logical level, we may understand loss of appetite as a pseudo-solution to communicative problems between the developing personality and the domestic environment of the overprotective home. At the social level, anorexia is a sick role which provides 'solutions' to the demands of a competitive middle-class culture through the secondary gains of the sick role. At the societal level, it is an effect of fashions relating to food, consumption and life-style. Anorexia is peculiarly expressive of the personal and social dilemmas of educated, middle-class women, because it articulates various aspects of their powerlessness within an environment that also demands their competitive success. In rather conventional sociological language, anorexia is located at the intersection of these achievement and ascriptive dimensions. Because the condition is overdetermined, no monodisciplinary approach to either diagnosis or therapy will ultimately prove adequate.

The diagnostic and therapeutic context is further compounded by the fact that anorexia as a disease is evolving over time. Over three decades of clinical experience, Bruch noticed that the disease had become more widespread through the social class structure, embraced a wider age group and was more prevalent (Crisp *et al.* 1976). It is estimated that between 5 and 10 per cent of the American adolescent female population are affected. Despite these difficulties, psychiatry now has a much clearer understanding of anorexia nervosa as a condition. For example, it is now evident that anorexia nervosa was in fact a misnomer, since the condition involves as much control over appetite as its loss. The classic anorexic is the master (the gender specificity of 'to master' is another complexity) of disguise and deception, only acting as if she had no appetite, while being fascinated by food.

CONCLUSION

Despite these improvements in medical understanding, the pro-spects for the management of eating disorders are bleak. Eating and dieting are of course the topics of a global agro-industry which, through advertising symbolism, connects food with personal status, sexuality and sociability. Obesity is stigmatized as a sign of moral weakness (Cahnman 1968). Because the transition from adolescence to adulthood has become more rather than less complex in a post-industrial society, it is likely that eating disorders will increase rather than decrease. These sociocultural problems are compounded for women because we have become an 'obesophobic society'. Although fatness has become a stigmatic sign which transcends class and gender, the demands of ascetic thinness weigh more heavily on the bodies of women than of men. Thus, the bio-politics of the late-twentieth century may well provide a new and sinister meaning to the nineteenth-century slogan – 'man is what he eats'.

In this discussion of Hilda Bruch's approach to anorexia nervosa, I have attempted to draw attention, within a general discussion of the recent social science literature, to the value of a phenomenology of the mouth as a clue to anorexic behaviour. To have a mouth is to have the possibility of an intentional linkage between *Umwelt* (the biological environment) and *Welt* (cultural world of meaningful action). Loss of appetite is phenomenologically parallel to loss of speech, and both conditions point to the absence of social voice, which permits us easeful as opposed to diseaseful communication. In turn, anorexia as disease celebrates our ambivalent location between *Welt* and *Umwelt*. These sociological enquiries into health and illness lead inevitably therefore into philosophical enquiries into our ontological status between culture and nature (Gehlen 1988), which in turn underlines the claim that the future development of sociology (especially in relation to disease categories) will require a more fully developed sociology of the body.

REFERENCES

Abercrombie, N., S. Hill and B.S. Turner (1986) *Sovereign Individuals of Capitalism*, London: Allen & Unwin.

Alexander, F.G. and S.T. Selesnick (1966) *The History of Psychiatry*, New York: Menter.

Bell, R.M. (1985) *Holy Anorexia*, Chicago and London: Chicago University Press.

Boyne, R. (1990) *Foucault and Derrida, the Other Side of Reason*, London: Unwin Hyman.

Brown, P. (1981) *The Cult of the Saints, its Rise and Function in Latin Christianity*, Chicago: Chicago University Press.

Bruch, H. (1957) *The Importance of Overweight*, New York: Basic Books.

Bruch, H. (1973) *Eating Disorders: Obesity, Anorexia and the Person Within*, New York: Basic Books.

Bruch, H. (1974) *Learning Psychotherapy, Rational and Ground Rules*, Cambridge, Mass: Harvard University Press.

Bruch, H. (1977) *The Golden Cage. The Enigma of Anorexia Nervosa*, Cambridge, Mass: Harvard University Press.

Bruch H. (1988) *Conversations with Anorexics*, New York: Basic Books.

Brumberg, J.J. (1988) *Fasting Girls, the Emergence of Anorexia Nervosa as a Modern Disease*, Cambridge, Mass: Harvard University Press.

Bury, M.R. (1986) 'Social constructionism and the development of medical sociology', *Sociology of Health and Illness* 8, 2, 137–69.

Bynum, C.W. (1984) 'Women mystics and eucharistic devotion in the thirteenth century', *Women's Studies* vol. 2 (1–2):179–214.

Bynum, C.W. (1985) 'Fast, feast and flesh: the religious significance of food to medieval women', *Representations* vol. 11:1–25.

Bynum, C.W. (1988) 'Holy anorexia in modern Portugal', in *Culture, Medicine and Psychiatry* vol. 101, 12, 2, 239–48.

Cahnman, W.J. (1968) 'The stigma of obesity', *The Sociological Quarterly* vol. 9(3):283–99.

Caskey, N. (1986) 'Interpreting anorexia nervosa', pp. 175–208 in S.R. Suleiman (ed.) *The Female Body in Western Culture*, Cambridge: Cambridge University Press.

Cohen, R.S. and T. Schnelle (eds) (1986) *Cognition and Fact, Materials on Ludwik Fleck*, Dordrecht: D. Reidel Publishing Co.

Crisp, A.H., R.L. Palmer and R.S. Kalucy (1976) 'How common is anorexia nervosa? A prevalence study', *British Journal of Psychiatry* vol. 128:549–54.

Davies, M. (1982) 'Corsets and conception: fashion and demographic trends in the nineteenth century', *Comparative Studies in Society and History* vol. 24:611–41.

Eckerman, L. (1987) 'Selfhood versus sainthood: towards a social conception of anorexia nervosa', pp. 57–64 in S. Abraham and D. Llewellyn-Jones (eds) *Eating Disorders and Disordered Eating*, Sydney: Ashwood House.

Escamilla, R.F. (1944) 'Anorexia nervosa or Simmonds' Disease? Notes on clinical management with some points of differentiation between the two conditions', *Journal of Nervous and Mental Disorders* vol. 99:583–7.

Feher, M., R. Naddaff and N. Tazi (1989) *Fragments for a History of the Human Body, Part Three*, New York: Zone.

Feighner, J.P. (1972) 'Diagnostic criteria for use in psychiatric research', *Archives of General Psychiatry* vol. 26:57–63.

Foucault, M. (1967) *Madness and Civilization*, London: Tavistock.

Foucault, M. (1987) *The History of Sexuality; Volume 2: The Use of Pleasure*, Harmondsworth: Penguin Books.

Fox, R.A. (1976) *The Tangled Chain, the Structure of Disorder in the Anatomy of Melancholy*, Berkeley, Los Angeles: California University Press.

Gehlen, A. (1988) *Man, his Nature and Place in the World*, New York: Columbia University Press.

Gerhardt, U. (1989) *Ideas about Illness, a Medical and Political History*, London: Macmillan.

Hayes, D. and C.E. Ross (1987) 'Concern with appearance, health beliefs and eating habits', *Journal of Health and Social Behaviour* vol. 28(2):120–30.

Holton, R.J. and B.S. Turner (1986) *Talcott Parsons on Economy and Society*, London: Routledge & Kegan Paul.

Lawrence, M. (1979) 'Anorexia nervosa – the control paradox', *Women's Studies International Quarterly* vol. 2:93–101.

Lidz, T. (1985) 'In memoriam: Hilde Bruch M.D. (1904–1984)', *American Journal of Psychiatry* vol. 142:869–70.

Loudon, I. (1980) 'Chlorosis, anaemia, and anorexia nervosa', *British Medical Journal* vol. 281:1669–75.

Lyons, B.G. (1971) *Voices of Melancholy, Studies in Literary Treatments of Melancholy in Renaissance England*, London: Routledge & Kegan Paul.

McFarland R.E. (1975) 'The Rhetoric of Medicine: Lord Herbert's and Thomas Carew's Poems of Green Sickness', *Journal of the History of Medicine* vol. 30:250–8.

MacLeod, S. (1981) *The Art of Starvation*, London: Virago.

Merleau-Ponty, M. (1962) *The Phenomenology of Perception*, London: Routledge & Kegan Paul.

Nietzsche, F. (1974) *The Gay Science*, New York: Viking.

Palmer, R.L. (1979) 'The dietary chaos syndrome – a useful new term', *British Journal of Medical Psychology* vol. 52:187–90.

Palmer, R.L. (1980) *Anorexia Nervosa*, Harmondsworth: Penguin Books.

Parsons, T. (1951) *The Social System*, London: Routledge & Kegan Paul.

Pina-Cabral, J. de (1986) *Sons of Adam, Daughter of Eve, the Peasant View of the Alto Minho*, Oxford: Oxford University Press.

Prince, R. (1985) 'The concept of culture bound syndromes: anorexia nervosa and brain-fag', *Social Science Medicine* vol. 21(2):197–203.

Siddall, A.C. (1982) 'Chlorosis-etiology reconsidered', *Bulletin of the History of Medicine* vol. 56:254–60.

Sontag, S. (1978) *Illness as Metaphor*, New York: Farrar, Straus and Giroux.

Sontag, S. (1989) AIDS and its Metaphors, New York: Farrar, Straus and Giroux.

Swaan, A. de (1981) 'The politics of agoraphobia', *Theory and Society* vol. 10:359–85.

Turner, B.S. (1982) 'The discourse of diet', *Theory, Culture & Society* vol. 1 (1):23–32.

Turner, B.S. (1984) *The Body and Society, Explorations in Social Theory*, Oxford: Basil Blackwell.

Turner, B.S. (1987a) 'Agency and structure in the sociology of sickness', *Psychiatric Medicine* vol. 5(1):29–37.

Turner, B.S. (1987b) *Medical Power and Social Knowledge*, London: Sage.

Varma, S. (1979) 'Anorexia nervosa in developing countries', *Transcultural Psychiatric Research Review* vol. 16 (April):114–15.

Conclusion
Theory and epistemology of the body:
an interview with Richard Fardon

RF Over the last decade your work has ranged widely – both theoretically and empirically – and you have written a great deal. Has there been an overall plan or goal?

BT There is always a danger in retrospectively constructing a plan for one's work, but certainly I've wanted there to be an overall consistency and coherence to the material I have written. The more I've written the more problematic the idea of a plan or a goal has become, especially as I have worked in a number of different fields, particularly sociology of religion, medical sociology and sociological theory. I think there is a plan or structure which comes from the original work in *For Weber* (Turner 1981). I started from Weber's demand that the task of sociology is to understand the characteristic uniqueness of the times in which we live, and I thought that Weber's main idea about that 'characteristic uniqueness' was contained in his understanding of the process of rationalization. I was very interested in the sociology of religion when I started reading Max Weber seriously, therefore, the ideas of secularization and rationalization framed the theoretical issues on which I launched out.

   As a result I wrote three interrelated books. *Religion and Social Theory* (Turner 1983) was a study of the relationship between the body, theodicy and religion. *The Body and Society* (Turner 1984) was, as the subtitle suggests ('explorations in social theory') an attempt to spell out some of the theoretical problems in a study of the body. *Medical Power and Social Knowledge* (Turner 1987) attempted to apply the results of my theoretical work to a substantive topic – the place of medical practices in modern society. There was a fourth book, which goes somewhat beyond the specific topic of our discussion, *Nietzsche's Dance* (Stauth and Turner 1988). This was written with Georg Stauth to examine the theme of the body in Nietzsche and trace the influence of his philosophy on Weber, Freud and Foucault.

Religion and Social Theory, the first volume of what I see as a trilogy on the body, was written while I was at the University of Aberdeen. This was the beginning of an attempt to think about the body and theodicy as the basis of the work I wanted to do. When I came to write *The Body and Society*, much of the groundwork had been prepared for elaborating the idea that one aspect of modernization was the progressive management, surveillance and regulation of the human body, of which sexuality might be a particularly prominent feature. Questions about women and sexuality dominated *The Body and Society*. The work I was doing on medical sociology followed easily from the first two books. *Medical Power and Social Knowledge* was partly based on lectures that I gave at Flinders University in South Australia. By the time I wrote this book, I had become much more influenced by the work of Michel Foucault. The book on medical power employs a framework in which Foucault and Weber are integrated and, although it is in some ways a textbook, it does have an underlying theme which concerns the transfer of social control from religion to medicine.

RF In *Medical Power and Social Knowledge* you write that you see the institutions of medicine, religion and law performing interchangeable functions at different times (Turner 1987:19). So, by adding that third institutional complex, we could see significant continuities in your interests that would include works other than the trilogy on the body – for instance, your accounts of equality and citizenship (Turner 1986a; Turner 1986b). But, to leave aside the continuities for a second, to what extent have your theoretical perspectives changed in writing the trilogy? When we first met around 1981, when you were writing *Religion and Social Theory*, you described the approach you wanted to take as 'materialist': particularly concerned with the regulation of bodies and the regulation of populations. In the most recent book of the trilogy, you say your aim is to integrate three levels of analysis which you call individual, social and societal (Turner 1987:5). If you take us through the theoretical development of the trilogy, then we could go on to talk about the changing relationship between your readings of Weber and Foucault.

BT I did my postgraduate work on the sociology of religion with special reference to the nature of religious commitment. As a consequence of this interest I became increasingly influenced by Max Weber. When I started my academic career I was committed to developing sociology within a Weberian tradition. But in Britain in the late sixties and early seventies, much of the theoretical centre ground was occcupied by Althusserian Marxism, and it was difficult to avoid an engagement with the legacy of Marx, and with structuralist Marxism in particular. While I was teaching in the Department of Sociology at Lancaster in the mid-seventies, the Althusserian paradigm and the work of Nicos Poulantzas were very prominent, both in undergraduate teaching and in postgraduate research and publications. Therefore I was pulled, quite willingly, into the debate between Weber and Marx. I've always wanted to argue that one doesn't really need to choose between them because there was such an overlap between their work on alienation and rationalization. Some of these arguments appeared in *For Weber* (Turner 1981) and most recently in *Max Weber on Economy and Society* (Holton and Turner 1989) and *Max Weber, from History to Modernity* (Turner 1992). To suggest that there are two paradigms, and that one has to choose one rather than the other, seems to me to be completely misplaced. I wrote *For Weber* with an obviously ironic title: *For Weber* was a response to *For Marx* (Althusser 1969). When I was talking about materialism in the early work it was very much within a paradigm influenced by Weber. I wanted to argue that Weber's interest in violence for example had to be seen as compatible with Marxist concerns with economic control and regulation. Whereas Marx concentrated on the monopoly of economic force, Weber also took account of monopolies of physical and spiritual violence. I took the position that Weber provided an important supplement to the idea of economic determinism. I would approach the whole issue rather differently now, but I think the arguments of *Weber and Islam* (Turner 1974) and *For Weber* are still broadly correct.

The subtitle of *Religion and Social Theory* was 'A materialist perspective', and there was obviously an ironic note in this also, because it was taken from an interview with Foucault. When he was asked about the problem of ideology in Marxism, Foucault (1980:58) replied 'I wonder whether, before one poses the question of ideology, it wouldn't be more materialistic to study first the question of the body and the effects of power on it'. While I found Foucault's work very seductive, studying Foucault actually took me back to re-reading the early Marx and Marx's critique of Feuerbach; it led me to an interest in both philosophical anthropology and Marx's notion of human practice. My attempts to use ideas about materialism, sensual praxis and the dialectic of nature, need and culture were a consequence of that particular reading of Foucault and Althusser. When I came to write *The Body and Society* I was trying to extend and elaborate that materialistic paradigm to take on board the early Marx and the debates about Marx's view of nature and human nature. Incidentally, these issues were quite prominent in Australian anthropology and sociology at the time, where the so-called Hungarian circle, which included Agnes Heller at La Trobe, was influential. I was also increasingly influenced by debates in feminist theory, which partly explains why the theory of patriarchy was so prominent in *The Body and Society*. I was trying, as it were, to provide a materialistic foundation to social science in a reworking of the whole debate about biology, nature and culture which was the legacy of a certain type of Marxism. This foundation would be bound up – *not* with the crude notion of economic production – but with a philosophical anthropological idea of human practice that was grounded in a notion of anthropological need or a Marxist ontology.

By the time I came to write *Medical Power and Social Knowledge* much of the Marxist paradigm had either dropped away or been taken for granted. By that time I was much more significantly influenced by continental philosophy and in particular by the legacy of Foucault. It seems to me that Foucault provided a very substantial framework for trying to recast much of the debate in medical sociology about control, surveillance and the body. In the final book of the trilogy, much of the Marxist paradigm has disappeared apart from the idea that economic processes and the whole structure of society are fundamentally bound up with the problems of health and illness, for example in Third World societies, as I try to show towards the end of that book. One indication of the changing emphasis in my work is that in the new edition of *Religion and Social Theory* (Turner 1991a), I have abandoned the subtitle ('a materialist perspective') as no longer an important specification of the main title.

RF   Let's talk about the relationship between your readings of Weber and Foucault, especially what is, I think, your changing reading of Weber. I'll give you a summary and ask if you agree with it. The subtitle to the collected essays on Weber, 'the sociology of fate', indicates that your early reading of Weber was less attracted to the interpretative aspects of Weber's theory of social action than to his ideas of rationalization and unintended consequence which you summarize as fatalism. This interest must have converged, in the light of what you have already said, with the thrust of structuralist Marxism – that men do not make history under conditions of their choosing. Your early reading of Foucault, it seems to me, was as a yet more extreme version of your fatalistic Weber. But this changed your reading of Weber, since you subjected the Foucault of control, surveillance, discipline and panopticism to a sociological critique from a Weberian perspective, for instance, in the paper published in a volume I edited (Turner 1985). Reading Foucault seems to have led you to recuperate a different reading of Weber.

BT   I agree that the early approach to Weber was developed in the context of structuralist criticism, and *For Weber* was largely a reply to the Poulantzian criticism of Weberians as subjectivist, humanist and individualist. I wanted to provide a structuralist version of Weber to stand alongside or to replace the structuralist version of Marx. I was, by the way, very much influenced by Fredric Jameson's interpretation (1973) of the narrative structure of Weber's sociological theory. I wanted to object to the idea that Marxism was objective and scientific, whereas Weberian sociology was just bourgeois subjectivism. I tended to focus on things like Weber's emphasis on militarism, on the institutions of violence, on the unintended consequences of action, and on Weber's economic sociology. The idea of fatalism was that whatever human beings might want or desire, they wouldn't necessarily get them, or they would get them in a way they hadn't anticipated. At the time, I saw myself presenting a very deliberately structuralist reading of Weber which I think rather down-played both the idea of interpretative sociology and the problem of meaning in human action. In retrospect, I am far more conscious of the widespread impact of *Kulturpessimismus* on German philosophy and social theory. The concept of fate was not particularly unique to Weber, as is demonstrated in Liebersohn's wonderful study (1988) of *Fate and Utopia in German Sociology*.

When I read Foucault I initially saw him merely reproducing much of Weber's analysis of discipline and the penitentiary. In fact, I was amazed that Foucault never quoted Weber in his studies of discipline, punishment, the military and bureaucracy. I was impressed by the idea that Weber had already argued that the origins of modern ascetic discipline and rational organization were to be located in the monastery and the army, and that Foucault, in *Discipline and Punish*, seemed to choose exactly the same illustrations in order to get at his view of the carceral society. His analysis of the growing detailed character of surveillance and regulation suggested a close correspondence with Weber's views of bureaucratization. That interpretation of Foucault has not been acceptable to those defenders of Foucault who want to see Foucault as an essentially original and autonomous thinker.

RF This reading seems to privilege the work of the later Foucault, the writer of *Discipline and Punish* and volume one of *The History of Sexuality* rather than *The Order of Things* (1974a) or *The Archaeology of Knowledge* (1974b).

BT I think it was Paul Hirst who criticized me for providing a sociological reading of Foucault, and I admit that I tended to see Foucault within the paradigm of a sociology of discipline or a sociology of asceticism. I took less seriously the epistemological work on knowledge and discursive formations.

RF And now?

BT I think that my appreciation of Foucault is probably wider and deeper. I am sure I was too narrow in my criticism or reading of Foucault. He was a much more interesting writer than I originally suspected. I have become more aware, for example, of Foucault's work on the self (Martin *et al.* 1988). I realize that he produced work which stands outside and beyond the original Weber problematic with which I was working, Foucault made a major and original contribution to twentieth-century thought. However, interpretations of Weber have also changed over the last two decades. The work of Wilhelm Hennis (1988) shows that Weber was primarily concerned with the study of life-orders and personality, or what I have called 'characterology'.

RF I am struck by your preference for emphasizing convergence rather than divergence between Weber and Foucault. For instance, your early book *Weber and Islam* (Turner 1974), involved an examination of the comparative study of rationalization that Weber generalized from his intensive account of the Protestant Ethic. Coming more recently to Foucault's work on the technology of the self, you nevertheless find connections with the Protestant Ethic thesis. There seems to be a predilection for strongly convergent readings of the major theorists.

BT The way I've approached thinking about major theories in social theory might be called a 'strategy of inclusion' rather than a 'strategy of exclusion', and this approach reflects my disappointment about the absence of any significant cumulative theory building in sociology (Turner 1988). There is an unfortunate tendency in sociology where people adopt a particular theoretical tradition, or a social theorist, or a particular paradigm, and then proceed to destroy all the other traditions, theorists or paradigms as competitive or incompatible with their preferred position. The more one can destroy the better. This strategy of annihilation is closely related to the whole problem of masters and disciples, which seems to dominate a lot of sociological theory. Thus, if a person is committed to radical Marxism, they want necessarily to reject everything Parsons, Weber, Foucault or Durkheim ever wrote – and so on for other affiliations. People who operate in this exclusionary fashion take criticism to mean brutal destruction of alternative paradigms. This is the mentality of sectarianism. Of course, the idea that sociology is characterized by sectarian fundamentalism which precludes any possibility of synthesis and integration is hardly original (Robertson 1974:107).

By contrast, I've tried to provide a synthesis which is what I'm today calling a strategy of inclusion. This means that I have tried to extract what I've found to be useful or relevant in a range of writers, but in particular Weber and Foucault and, in medical sociology, Parsons. My approach is to think about problems that interest me from diverse starting points. For instance, it seemed odd to me not to take seriously the work of Erving Goffman when writing on the body, although in general I think that Goffman's work has been very adequately criticized by, for example, neo-Marxists like Alvin Gouldner. It also seemed impossible to write about the body without serious consideration of the work of Merleau-Ponty, whose paradigm is very different from the early Marx or Foucault. In selecting a range of topics that interest me, such as the body or politics or rationalization or religion, I aim to formulate a theoretical position which, rather than playing upon the wars between the various schools in social science, actually tries to draw some sort of peace between the competing paradigms in order to produce something which I think is literally more productive than this highly divisive critical strategy of exclusion. Too many introductory textbooks to sociology have been content merely to describe apparent incompatibilities between various approaches to the study of society.

RF Is it worth diverting for a minute to talk about the way in which you use the idea of paradigm competition in sociological theory? You talk about that at the end of *Medical Power and Social Knowledge*. Your view of paradigms in social theory, if I understand you correctly, is that they continuously compete rather than replacing one another. The plurality of paradigms is a necessary condition of sociology. Am I reading you right there?

BT I think so, except let me express it this way. What I'm trying to do in the medical book is to argue that different problems occur at different levels and some are more general than others. This is the assumption behind the description of the three levels of the individual, the social and the societal (Turner 1987:5). I think now I'd add some more levels which would be the organic, as a level below the phenomenological, and possibly a level beyond the societal which might be the ecological or environmental. My argument assumes that sociologists don't have to choose, as it were, between levels. For example, it seems perfectly sensible for a medical sociologist to be interested in the question of human pain, and I felt that the best approach to that was the work of Merleau-Ponty on phenomenology. However, it was equally sensible for sociologists to be interested in the interaction between doctors and patients, and I felt that the legacy of Parsons's 'sick role' (Parsons 1951) was the most appropriate approach to that level. Finally, I felt that neither Merleau-Ponty nor Parsons was adequate for understanding the impact of the state or the environment or social class on the distribution of health and illness globally. One had to look for a neo-Marxist or political economy paradigm to examine these questions. I think this is an illustration of what I intend by a 'strategy of inclusion', that, rather than being forced to choose between those particular competing paradigms, one could see them as addressing very different issues at rather different levels. Another way of describing this procedure is to talk about epistemological pragmatism, that is epistemological and theoretical questions should be framed in terms of specific levels or specific problems, depending upon the range of interest, the topic and orientation of the sociologist.

RF Now you're using two different vocabularies to describe what may be the same thing. In one vocabulary you're talking about paradigms with the implication that they are typically competitive or at least in sociological practice they tend to be. But you shift sometimes into a vocabulary of levels which has a different connotation because levels are appropriate to problems whereas paradigms are competitive. Do you see those two vocabularies as being quite distinct?

BT I think the first problem is one of professionalism, and we could look at it through the work of Pierre Bourdieu: a lot of the theoretical competition that goes on in any social or natural science is a function of competition over resources. Part of the exclusionary strategy is a consequence of struggles between élites to monopolize resources, including cultural capital.

However, to take your question head-on, there is obviously a tension, or possibly even a contradiction, in trying to argue either that paradigms don't need to be consistently or continuously replaced by each other, and arguing that we can think about a range of issues in social sciences operating at different levels. I think the problem with the inclusionary strategy revolves around the question as to whether my approach can be infinitely eclectic. I have been quite heavily criticized for theoretical eclecticism; I do not feel, however, that this is an adequate description of what I am trying to achieve. What I would try to defend now is the idea of epistemological pragmatism, that the theory, the paradigm and the approach may very well depend on the type of issue a theorist is interested in. Of course, the medical sociology book, in distinguishing levels, is making a commentary on 'structuration theory'. By keeping the levels apart, I am implying that there is no satisfactory or global solution to the agency/structure issue.

The question about levels also raises questions about interdisciplinarity. For example, to what extent are psychology or phenomenology particularly geared to dealing with individualistic problems about pain or suffering? Sociology might be that discipline which is more concerned with the institutions that surround health care. Economics refers to the problems of economizing behaviour relevant to health-care resources and so forth. I think that your question about the double problem of levels and paradigms also has to be discussed in the context of competing disciplines.

RF I think that in introducing interdisciplinarity you're multiplying the number of sites to which you want to apply the notion of level. Do these levels exist pre-paradigmatically? Are the levels given, so that you can then choose the paradigm? Or are the levels actually defined by your choice of paradigm?

BT Well, I can see that the way in which I formulate it might suggest that I am adopting naive empiricism: social science simply adopts a given set of problems that present themselves to the observer. Obviously, I don't want to adopt such a naive strategy of empiricism; there are no theory-independent facts (although some facts may be more theory-dependent than others). A Marxist paradigm picks out certain issues for research – class and health, the nature of production in relation to sickness, the relationship between welfare and economics. I do not deny this obvious lesson. Clearly, I want to adopt a position in which, at the very least, there is an interaction between the theorization of problems and the nature of the 'problems' which exist in public discourse. However, in actual research one is always faced with relatively mundane issues: what should I study and which paradigm appears most relevant here?

RF I still have problems with the notions of levels and paradigms, because some of the positions you wish to use inclusively were specifically proposed with an exclusionary purpose. Can you reconcile competing paradigms by appealing either to level or discipline? Think for instance of Foucault's relationship to phenomenology. Foucault intended his work to supersede phenomenology (if we believe Dreyfus and Rabinow 1986).

BT It seems to me that Foucault always rejected the idea that he could be easily labelled, and I've always adopted the approach that once a work has been published it is public property. It can be interpreted and used in ways which may contradict the author's original intentions. If I perceive a relationship between Foucault and Weber, there is no constraint to prevent developing a theory in which that reconciliation or integration of approaches actually happens, regardless of what Foucault might have thought of Weber or, to take your own illustration, regardless of what Foucault might have thought about Merleau-Ponty. I don't feel that one is bound by Foucault's particular criticisms of that phenomenology. In any case, Foucault's critique of phenomenology can only be understood in the context of political battles inside French intellectual circles, especially the conflicts between Marxism, existentialism and structuralism. Those struggles may not be very relevant to our context.

RF I accept that Foucault's writings need to be read against the political context of his intellectual world but, obviously, you would not want to reduce the issues at stake to political strategy. If you wanted to retain Foucauldian and phenomenological approaches as appropriate to distinct levels of analysis, would you have to demonstrate that Foucault was wrongly critical of phenomenology in some respects?

BT I don't think that one can adopt a cavalier position in which it really wouldn't matter whether Foucault was right or wrong. No, I think you're correct that I would have to elaborate a much more convincing argument about Foucault's views of phenomenology being, if not incorrect, at least contentious, if one wanted to achieve real analytical elegance. In principle, there is nothing especially contentious in the notion of levels of analysis. It is a familiar notion in macro-economics and micro-economics, for example. I think Parsons's heuristic use of the personality-society-culture division is unproblematic. Thus, my argument that it is pragmatically useful to distinguish between individual–social–societal is merely to say that typologies are heuristically useful.

RF I am less sanguine that we can judge between what is useful and what is merely convenient so easily. But let me try and put all this another way. You've been claimed as an eclectic, and you've accepted the designation subject to the redefinition that you really want to adopt an inclusionary strategy where the level of analysis responds to empirical problems that sociologists can address; in this way, sociologists may be able to say something that is going to be helpful and constructive to other people who may not be sociologists – such as nurses, doctors, health workers or whoever. The obvious question that is put to a theoretical eclectic is whether the theories that are being pressed into service are contradictory. Empiricism is then in danger of becoming a strategy of opportunistic movement between levels in relation to different problems rather than an effort to sort out the tensions, or even the contradictions, that might exist within the different theoretical frameworks that you are using.

I want to take a specific example of this in relation to sociological terminology. It's noticeable in your books that you employ a very catholic range of terminology, and you feel happy to use terms that are often semi-synonyms. Thus you use 'rationalization', 'modernization', 'secularization', 'surveillance' and so forth to describe aspects of what you presumably see as related areas of concern. You're willing to use terms like 'structure', 'function', 'system', 'society' but also 'power-knowledge', 'agency', and so forth. You don't seem to have a principled limitation on the range of terminology that you are willing and quite happy to use in pursuit of different problems. This might give the impression that your epistemological pragmatism, to use your phrase, is rather unprincipled.

But, against this view, I can also see a consistency between the books, and this comes out partly in the theoretical accretions elements that don't appear in more than one book. For instance, the notion of 'society', which is important in *The Body and Society*, disappears in the third volume of the trilogy where, presumably because you are looking at medical professions and institutions, your analytic categories are much less broad: particular types of medical institution, the clinic, the asylum, the professional body of medical practitioners and so on. This suggests to me that the elements of the eclectic

theoretical brew that stay with you longest are those that echo
your original reading of Weber. Thus, the levels you
distinguish in *Medical Power and Social Knowledge* seem to me
the same as the levels that are problematic in Weber's general
theory. Your phenomenological level corresponds to Weber's
discussion of the difference between meaning and behaviour;
your social level corresponds fairly closely to Weber's
discussion of the ideal types of authority, legitimacy, institutionalization and so forth; and the societal level, as you're using
it, corresponds to the master processes both in Weber and
Foucault. The societal level is about secularization, rationalization, surveillance, disciplinary practices and panopticism. So,
as devil's advocate to a self-confessed theoretical eclectic and
epistemological pragmatist, I wonder whether behind the
appearance of widespread borrowing there isn't a long-
term continuity in your Weberian proclivities supplemented
by helpful and sometimes short-term accretions from elsewhere.

BT Let me try to answer that very important question in a very
broad way. And I'll start again with the context within which
all of this has been written. Sociology, particularly in Europe,
is often criticized for its internal confusions and conflicts, for
competing paradigms, and for fashionableness in perspectives
and approaches. The paradigmatic sectarianism is reflected,
according to its critics, in endless jargon. I think this criticism
is often misplaced, but one has to admit that throughout much
of the sixties, seventies and early eighties there have been waves
of paradigms with enormous shifts from Marxism, to symbolic
interactionism, to ethnomethodology, to neo-functionalism,
to neo-Marxism, to structuralism, to poststructuralism and
so forth. Trying to teach sociology and do research in sociology
within this context has been both politically and professionally
very difficult. In recent years there has been increasing
discusssion about the possibility of a new synthesis, of there
being an end to the war in paradigms. To give you an
illustration, I think the work of Jeffrey Alexander (1987) in
neo-functionalism, and his attempt to build a 'multi-
dimensional theory', has been concerned to put an end to the
ceaseless round of sectarian violence that we have had in
theoretical debates in sociology. And I think I've always had
a strong commitment to the idea of sociology as a serious,

unified, important and relevant practice. For good or for bad, I was very committed to the idea of a classical sociological tradition that was still relevant, and that one had to work with rather than against that tradition. My theoretical inclusionary strategy has to be seen in the context of wishing there to be an end to the endless sectarian confrontation in social theory.

You're correct to suggest that there are limits to my own epistemological pragmatism and theoretical eclecticism. It's certainly the case that throughout this work on the sociology of the body the ideas of Weber in relation to concepts like theodicy, secularization and rationalization are important. Also Weber's ideas about action, interpretation, meaning and so forth have played an important part. I think it's for that reason that I've chosen the topics of religion, law and medicine as particular sites from which to look at the body from the point of view of a Weberian paradigm. If you look at the sort of theories that I include, I think you are probably right to suggest that there is a limit to the number of actual theoretical positions in my work. Although Parsons, Foucault and Weber have been important to the three books, in practice much of the theoretical drive comes from Weber. I've added Foucault's debate about surveillance and discipline, and I've adopted or adapted certain aspects of Parsons's views on the four functional problems of any social system and his specific analysis of the sick role.

To come to the more particular aspect of your question, I think it's interesting that you pick upon the idea of 'society' as something that might have changed over the three books. Certainly, in the first book on religion there is a Durkheimian view of society in relation to the body, and I tried there to bring out some linkage between the body, religion and society in terms of problems about social cohesion and social solidarity, but there was also an obvious Weberian aspect, namely the problem of meaning. In the second volume, the word 'society' appears in the title of the book and, as I've already discussed, much of that influence is from early Marx and Foucault. In the final book on medical sociology, you're correct to suggest there is a much greater emphasis on the idea of institutions. One of the background problems to the book was the political collapse or erosion of organized communism. At the end of that book I conclude by arguing provocatively that, regardless of the type of social system or society, there would be a range of common medical/social problems related to fundamentals such as birth, maturation, death, pain and disease. I point out that, regardless of the mode of production or society, the range of killer diseases in New York and Peking may be of a different order but basically people die from three major killers. There is a common set of processes in modern society which are global, and it may be for that reason that, rather than taking a society by society set of comparisons, I'm looking at more basic fundamental processes in the nature of ageing, disease, sickness and so forth. Therefore, the disappearance of the concept of society may be less to do with fundamental epistemological and theoretical issues and again more to do with the range of topics that I chose to look at in the medical book.

RF Let me come back on one or two things that you've said. I'm not wholly convinced by your disarming account of the concept of 'society' disappearing in relation to your chosen range of problems. It seems rather that your use of 'society' in the middle book is actually incompatible with the issues you address in the third. It seems inappropriate to the point you want to make about processes that have a global threat. You refer to a 'societal level' which addresses global issues of secularization, modernization, rationalization and so forth. These themes appear in both Foucault and Weber as particular story lines about the development of the West, and we ought to come back later to the point that your theoretical 'inclusionary strategy' is mirrored in an empirical inclusion of Western and non-Western places, in your example New York and Peking, in a single process at the societal level. At this level, it's difficult to see what role an idea of 'society' might play in illuminating these common processes.

To put it methodologically, your theorizing tends to be middle range (that is to say, you neither want to be a grand theorist, who is going to defend a particular version of the world against all-comers, nor an empirical sociologist, with your nose close to the local ground like an ethnographer). You're looking to middle-range theorizing to allow you to discuss a range of problems with the body that can be thought about sociologically. In terms of this middle-range focus, it is very difficult to know what empirical entity in our contemporary world 'society' describes.

BT This whole question of the concept of 'society' is obviously fundamental to sociology, and there are a number of conventional responses and answers to it. From a sociologist's point of view, the conception of the 'social' is the critical issue and you may be correct in suggesting that the idea of the 'society' becomes increasingly redundant as I try to work my way through the problems posed to sociology by the body, nature, culture and so forth. In addition, we are working in an intellectual context where postmodern theorists such as Jean Baudrillard have challenged the whole idea of the 'social'. What I wanted to say was that neither Foucault nor Weber, who have been influential in my development, has a strong sense of the importance of the idea of 'a' or 'the' society but tends to work with ideas about social processes, social dimensions, and social institutions. Indeed, Weber's concern with the sociology of action and the nature of the social has been a much more significant influence on my work than more positivistic or empirical notions of 'society'. I think that the concept of 'the society' in Parsons's work is probably less significant than is normally assumed because Parsons himself starts from a theory of the structure of social action; it was only in his middle period, when he was talking about 'the social system', that you find a change in vocabulary from social action to a concept of society.

RF Let's move to the sociology of action through a discussion of agency. I want to revert to your metaphor of levels. There are two separate levels at which you want to work in order to address the problems that are on any relevant theoretical agenda now. The top level concerns broad-range processes for which you're using the vocabulary of surveillance, rationalization, modernization, and secularization, which cannot be narrowly located in any particular society or nation state, but are global in scope. In the basement, to stay with the architectural metaphor, you want also to address the materiality of the individual body. Part of the problem that you face is filling in the intervening storeys of the building you are erecting.

In the second volume of the trilogy we've been discussing, 'society' appears as a convenient counter to the idea of the body; you're really interested in the body but need something to play it against. 'Society' seems to play a strategic role in your own thought, albeit in a rather short-term way. The four-cell model of society is presented in a straightforwardly functionalist framework in *The Body and Society* (1984:91); 'society' is seen to respond to various needs, or requirements of human existence which are pre-cultural, pre-social and therefore natural. It is not clear how this position is reconcilable with your realization that nature is a culturally contrived idea. Nevertheless, the idea of fundamental needs is important to your account in which society is represented as an agent capable of fulfilling needs which human bodies present, as it were, existentially or pre-socially and which must be satisfied in some way or another.

This is not a position you retain very long, because it can evidently be subjected to criticism. Your later understanding of agency, in writings explicitly on the topic, suggests that agency refers to human agents, and this takes us back to the question of social action. So could I start by asking you whether you want to restrict the concept of agency to human agency or if you see agency as a broader concept of which human agency is one particular example.

BT   Before we discuss agency, I do not see my work as a form of conventional functionalism. I have defended Parsons (Robertson and Turner 1991) not because I am a functionalist, but simply because I thought that Parsons had been wilfully misunderstood. The four-cell outline of *The Body and Society* was primarily a heuristic tool to organize the structure of the book. The peculiarities of this 'functionalist' strategy have been well understood and criticized by Arthur Frank (1991).

However, the broader question is fundamental to all social sciences: what is the relation between human agents and social structure? Any number of social theorists have attempted to come to terms with this problem. In Britain most recently Anthony Giddens's concept of 'structuration' (1984) has been a major attempt to resolve this problem, but earlier attempts by Parsons, Weber, Durkheim and Simmel took distinctive positions on this debate.

Certainly, my work has been based on a Weberian notion of social action, and I've also been trying to argue that the body is fundamental to any theory of action. One of the great weaknesses of modern analytical attempts to resolve this problem is that the body is either normally bracketed out or becomes part of the environmental conditions under which action takes place. When we talk about action theory in most conventional forms of sociology, we are usually talking about cognitive behaviour. In bringing the body back into sociology, I am attempting to find some theoretical space for dealing with questions of embodiment, and with questions about affect, ageing, sexuality and emotions in order to provide a more complete or adequate theory of action. It's true that there is a hiatus, or gap, between wanting, on the one hand, to talk about the embodied nature of the actor and, on the other, being interested in a range of macro-level questions in medical sociology. The links between these two levels are made via the ideas of production and reproduction. At least that was how I tried to think about the problem in *Religion and Social Theory* and *The Body and Society*, where I linked together the human actor as an embodied person and these broad historical structural constraints, which are very fundamental processes of demography; the production/reproduction of bodies and populations over time was one way in which I tried to link the micro-embodied actor with broad historical processes of the management of populations. In Foucault, I feel that this problem of what he called 'the accumulation of men' provides a theoretical strategy for thinking about how populations over the last three or four hundred years have been regulated, managed, understood and controlled in urban spaces by modern technologies. This provided a way of rethinking some quite conventional issues to do with demographic explosions and population changes, ageing and the greying of populations, the changing balance of the sexes and so forth. I realize that I utilize Foucault in a way many commentators might not accept: it is claimed that when Foucault talked about the concept of population he was talking about a way in which discursive formations are organized around a specific concept. I use the concept of population in a much more mundane, empirical, everyday sense of literally the accumulation of bodies in space presenting problems of social organization and social regulation. However, I think Foucault's appeal to 'population' is ambiguous.

RF To go back to agency – one of the things theory does, even implicit theory if we don't imagine ourselves working with explicit grand theory, is to identify certain features of our analysis as agentive; for instance, in the Marxist case class becomes agentive. Because we attribute effects to it, it is made agentive of other aspects of the analysis that are rendered – to borrow a concept from a colleague of mine (Mark Hobart) that might be conducive to you – patientive. Other elements become the patients of the particular agency acting upon them. This is why I asked you whether you thought that agency particularly referred to human agents, and to humans acting with an individualistic theory of social action. The functionalist position you took in *The Body and Society* had the effect of making society an agency which somehow responded to organic or pre-cultural needs as you then defined them. Is a theory of agency necessarily a theory of human agency?

BT   I don't think that a theory of agency is necessarily about human agency, and I don't see any problem in conceptualizing social classes or organizations or even societies as exercising agency. However, in writing about the body, my strong theory of agency must be a human theory; looking at my work more generally, for example, *The Dominant Ideology Thesis* (Abercrombie *et al.* 1980) or the little book I wrote on *Equality* (Turner 1986a), the agency of the human agent is a central feature. These studies allotted a major role to the place of human activity in the world; in other words, most of my work is not very structuralist but relies rather on a Weberian notion of the agent.

RF   You moved between the terms 'agency' and 'activity'. Is human activity human agency? What constitutes agency?

BT   When I talk about agency, I mean practices or actions that have an effect and bring about changes. So agency is the capacity of knowledgeable embodied human agents to bring about changes. That's leaving to one side the whole debate about unintended consequences and structures.

RF   In the final book of the trilogy you talk about agency in terms of choice, which presumably implies intention and a theory of rationality.

BT   This is the whole legacy of Weberian social action theory, and what I've added to that is the idea that the human body can't be regarded as part of the environment within which that agency takes place. My notion of the effective human actor includes the idea of the embodied human actor, whereas most other theories of agency tend to place the emphasis on the knowledgeable, conscious agent choosing goals in terms of norms and values. I would argue that this is based upon a false Cartesian paradigm and, like a lot of other people, I have criticized this legacy to expand the idea of agency to the embodied agent.

RF It seems to me that there are a number of theories of agency in your work, and it's not altogether clear which of them is compatible with other features of your eclectic theory. If we look at your writing on ageing and the reciprocity-maturation curve (Turner 1987:123), it seems that you use a transactional model of agency in which conscious actors presumably are aware of the balance of costs and benefits that are accruing to decisions they make about their actions. But when you talk about embodiment being a condition of human agency, I suppose you want to make a more than rhetorical statement. It's one thing to suggest critically that the body has been a neglected topic in sociology, but another to show exactly how embodiment is going to make a difference to theories of agency. How are we going to get beyond a rather simplistic transactional position of rational maximizing subjects, which of course corresponds to one of the Weberian models of rational action?

BT In talking about the embodied agent, I don't want to propose a version of methodological individualism; it is not possible to study the embodied human agent in isolation. Human agents are always embodied in some social context. The human body has inscribed on it millions of years of social development. By saying that the agent is embodied I am, at one level, merely indicating certain areas of research that become important once one has given some emphasis to embodiment. For example, there isn't an adequate sociology of pain, and it's only recently that sociologists have turned seriously to the problems of the sociology of emotions. There is lots of work on the representation, imagery and discourse of sex but not, from my point of view, a good sociology of embodiment *vis-à-vis* such issues as death, pain, sex and childbirth. By talking about the embodied actor I am indicating certain neglected areas that should come into greater prominence, or bringing into prominence work already done on embodiment.

The maturation-reciprocity curve is a version of exchange theory, but Durkheimian rather than utilitarian exchange. The argument is that, in collective terms, reciprocity is rewarded by symbolical and material social status. As people mature there is a tendency for reciprocity to decline and with it social status; the argument follows from those of *The Division of Labour in Society* (Durkheim 1960). It isn't closely related to the argument about embodiment, but I think Featherstone and Hepworth (1991) have demonstrated the complex relationship between the social construction of age, the social consequences of ageing and transformations of personal identity in their idea of 'the mask of ageing'.

RF Let's take the example of sexuality. You say that we already have a sociology that looks at the discourse of sexuality. How would it be different if it took the issue of embodiment seriously? Could you spell this out?

BT There isn't much research on the impact of ageing on sexual activity, sexual practice and sexual needs. One difference, for example, would be to take seriously the changing nature of the human body in relationship to personal identity and social performance. This would make a real difference to the way we do research and the topics we look at. I supervised a doctoral thesis in Australia which was partly concerned with the decision-making process leading to a mastectomy operation and showed how the medical model failed to understand the relationship between personal and sexual identity, and embodiment. In many other areas of medical sociology, a more self-conscious conceptualization of embodiment would expand our understanding of this interface between body, self-identity and social process – especially in the area of disability and impairment. The sociology of the body can make a powerful critique of the medical model.

RF To what extent does your judgement of the relative absence of a sociology of the body in disciplinary sociology also go for disciplinary anthropology? In your review (Turner 1991b) of recent trends in *The Body* (Featherstone *et al.* 1991) you suggested that the trajectories of interest in anthropology and sociology have been distinct in various ways.

BT Yes, I felt that *The Body and Society* in writing to sociologists (and perhaps historians and psychologists) had neglected a great, to use a word, corpus of knowledge in anthropology. I made some speculative comments on why the prominence of the body in anthropology contrasts quite significantly with its absence in sociology. I think anthropology has been particularly concerned with the body as a classificatory device and, while in contemporary Western urban societies the body has some function as a classificatory device, it's always much more obscured, masked and marginalized than is the case for pre-literate, tribal or agrarian societies. In the transition to urban, industrial, secular society this prominence of the body as a classificatory device becomes masked by alternative ways of classification. But there is a difference between the sociology of the body and anthropology because, through the work of people like Mary Douglas, anthropologists appear to be interested in the body as a means of discourse. They talk about it as a means of thinking about the world, whereas I think we need also to take into account the whole phenomenology of the body. This is another example of levels and disciplinary specialization. I do not deny the importance of studying the body as a classificatory principle; I merely say that there are other questions one might ask in anticipation of getting different answers.

RF   You're identifying anthropology with one tradition, the tradition of writing on the classification of the body which develops from the work of Marcel Mauss (1979) and of Robert Hertz (1960) in *Death and the Right Hand* and is continued in the Oxford School and notably Mary Douglas's work. I am sure that writers in this tradition would be unhappy to accept that bodily symbolism was unrelated to the phenomenology of the body; I think rather that the symbolic correspondences between the body and world (and vice versa) would be seen as a means to a phenomenological appreciation of the body. There's a lot of anthropological writing on other issues which might be of concern to a sociology of the body: for instance, on the aesthetics of the body concerned with body decoration and presentation, on bodily etiquette and so on. This constitutes a fairly coherent, and probably wrongly, separated body of anthropological writing. There is also a body of writing on emotions cross-culturally, stretching back at least to Malinowski's emendation of Freud to account for the matrilineal Trobriand Islanders, through American psychoanalytic approaches, to the 'personality and culture' school of Mead and Benedict, up to the present relativistic and non-foundational approach to emotions you find in the writings of Rodney Needham or Suzette Heald or Michelle

Rosaldo (Needham 1981; Heald 1989; Rosaldo 1980). These writers have emphasized the cultural specificity not only of the classification of human emotions but their reproduction through socialization, through ritual, and through the linguistic structuring of experience in different societies. I wonder whether this body of writing would be compatible with what I detected as a foundationalist or essentialist bottom line to your approach to the body as a natural fact?

You've agreed that there is a development in the trilogy on the body from a more materialist view of the body to a greater emphasis on a phenomenological level of analysis. But you're very unhappy about letting some foundational account of the body drop from your analysis, partly I suppose as a bridgehead back into medicine and other approaches to the body. Some recent anthropological literature has taken a phenomenological view of the body to its logical conclusion, which is to recognize that our own privileged discourses about the body, including the discourse of medical specialists, have to be seen as situated discourses and practices just as much as those of non-Western societies. Your argument that there is a universal experience of pain to which society or culture somehow responds would be difficult to maintain against a thorough-going culturalist approach. Indeed, it seems to me that you have moved a long way in this direction, for instance in writing about metaphorical bases of medical models of the body which change over time (see Turner on diet and the anatomy lesson in this volume). Do you think that you will be able to sustain your foundationalist account of the body against this other strand in your own thinking?

BT   The debate about classification, discourse, and the impact of poststructuralist methods of analysis is obviously difficult to ignore in sociology. In any case, I find some of these arguments very persuasive. We mentioned briefly the work of people like Emily Martin (before the interview), and I think in anthropology one can trace all of this back to Durkheim and Mauss (1963) on *Primitive Classification*. This literature normally involves an anti-foundationalist position. I believe that in this volume I have attempted to provide a general answer to this difficult question. In a nutshell, my answer is rather like Shakespeare's: I have yet to meet a philosopher who hadn't suffered from toothache.

However, to go back to the beginning of this discussion, it seems to me that there is a theme running through these volumes about theodicy; there are fundamental human experiences which are transcultural or universalistic and which point to a shared human ontology. The discussion of theodicy and morality in religious doctrine points to these fundamental human experiences, of which pain and suffering seem to me primary illustrations. I want to retain something of the legacy of that sociology of religion in my current discussion of the body; I want to try to hang on to some sort of foundationalist presupposition about common human experiences which are trans-social and trans-historical. One topic we might consider in context is ageing. On the one hand, it is evident that age in human society is socially constructed as a status position. The way in which human bodies are represented in terms of ageing processes has changed fundamentally. For example, in Western societies there has been a whole new emphasis on activity and on banning the concept of ageing as a decline of power. The image these days is of an endless youth stretching before us. But one knows that these representations of the young body can only be achieved either by continuous exercise and athleticism, topped up by the periodic face-lift, draining off human fat, operations to the eyelids and so on. These 'young' bodies are literally constructed, but they are constructed *against* ageing. The body is constructed by inter-vention, but we assume that there is something out there to be worked on. These fundamental questions about ageing and death, while socially constructed and changing across cultures, nevertheless have a foundationalist premise, if I can put it that way, which is this shared ontology of embodiment.

RF   Are you just saying that we get old and die? Does the shared ontology consist of more than the fact that human decay is inevitable?

BT   Most theoretical propositions can be reduced to nonsense; I obviously want to say more than we are born, we grow old and we die. I want to say that these events in both individual and social experience are deeply problematic because they are bound up with our personal identity, our place in society and the meaning we ascribe to ourselves. I want to argue the case for something like a shared ontology in order to be able to link the research I've done on the body with my other interests in questions about status, equality and citizenship. Sociologists, like anthropologists, find it difficult *not* to adopt a relativistic or culturally specific position on the question of ontology, but this epistemology of situated knowledges makes it difficult to deal with intersocietal connections or intercultural movements. It also makes political commitment difficult because, if we want to deal with questions about rights and obligations for example, we need a common language. In fact we need something that will replace natural law. A shared ontology or shared human experience makes it possible to mount an argument about things such as human rights. I am not convinced that a deconstructionist strategy can produce a viable politics in the long run.

RF But the discourse of human rights is a specific type of discourse, isn't it? A convention put together under particular institutional auspices – it isn't an ontological grounding of human experience but a specific form of discourse that has grown out of a specific tradition and certain sorts of institutions like the United Nations.

BT There is a specific human rights' discourse that has institutional horizons in a particular historical time and place, but if we adopt this stance – that it's only discourse, or only socially constructed – there is literally no foundation upon which one could have a shared position about the importance of rights. We would simply have a collection of preferences.

RF But aren't the rights *and* the agreement to them created discursively through institutions? I'm particularly curious why, having been a scholar of Islam and worked on non-Western societies, when you embarked on this project about the body you specifically restricted your curiosity, apart from a few excursions, to Western medical systems. It seems to me that this Western focus allows you to believe that there can be a shared ontology of the body – a notion that other parts of your work specifically dispute. For instance, you point out that medical models of the body are highly changeable, that they are metaphorically constructed from other areas of our lives, and that they change when our technology and production of ourselves change. Could I ask you, to put it in a rather concrete fashion, if you are suggesting that – Buddhists who believe in reincarnation, Africans who think that different elements of their body, say blood, bone and flesh, derive from different parts of their networks of kinship, New Guinea Highlanders who must continue to make payments to specific kin throughout their lives on account of the debts that arise from their embodiment because of the network of kin to whom they are indebted – all share a common ontology arising from something they are calling the 'body' which exists pre-culturally for them and can be recognized immediately and extra-texturally by us?

BT   A full answer to this question would have to take us into my changing relation with the debate on orientalism. However, there might be a fundamental difference between at least twentieth-century sociology and anthropology. The direction of anthropology is always towards cultural difference, suggesting there is no shared language, no shared experience, or shared bodies, only a proliferation of differences. Whereas the language of sociology tends in the opposite direction, which is a sort of language of sameness. Often this sameness has been argued from spurious theories such as convergence theory, suggesting that all societies are on some sort of evolutionary ladder towards an urban, literate, rational, secular society. Obviously, I don't believe in the idea of a common goal, teleology or historical process. The most superficial reading of contemporary history points to certain fundamental differences. On the other hand, the current debate about globalization is pointing towards the idea of a shared set of problems and, in some circumstances, shared cultures.

To come to your question, in the middle of the three-volume work on the body I became increasingly interested in philosophical anthropology, for instance the work of H. Plessner, A. Gehlen and F.J.J. Buytendijk, which points to a common ontology in the sense of a common set of human problems which you have reduced to the idea of being born, growing old and dying. The philosophical anthropological tradition does enable us to begin to talk about some shared experiences. While classificatory systems may be fundamentally different it may still be possible, from the point of view of philosophical anthropology, to talk about something like a shared experience of embodiment which, of course, is articulated in all sorts of different classificatory ways. My position is somewhat similar to that of Peter L. Berger who, building on the tradition of Gehlen, held a foundationalist view of human beings and had a constructionist epistemology.

RF As you say, nobody wants to deny that human beings must eventually die, but death does not exist apart from its significance. Is the significance attributed to death in different parts of the world somehow secondary to the fact of dying? Or, rather, is the fact of dying constituted in the understanding of what death is: whether I shall go to heaven or hell, whether I shall be reincarnated, whether I shall join my ancestors and so forth? Surely, this is not a secondary position, which comes after the fact that I'm going to die, but rather constitutes what the fact of my death will be.

I'm not trying to press you towards endorsing a strange anthropological, chopped-up world of totally discrete cultures. That seems to me to be a fiction of the way in which anthropology gets written; it doesn't correspond to any regionalized view of the way that, say, African cultures, are distributed. I do want to suggest that the datum of death, birth or ageing does not pre-exist the understanding that people have of it. I agree with you that quite how to arrive at cultural presuppositions, and to decide the degree and extent to which they are shared, are questions to which anthropologists have not sufficiently attended in the past; and this failure is open to a charge of obsession with difference that I accept. But your philosophical anthropological ontology suggests that biological facts pre-exist people's concern with them. This seems to me to move too far against difference. People are born into settings of sociality and culture which already have ways of problematizing and answering problems about death and ageing.

BT I don't think that anybody is born in a sort of cultural vacuum in which none of these meanings exist; quite the contrary, it's obvious that people are born into different cultures in which processes of sexuality, ageing, death, and dying are already schematized, elaborated and symbolized in cultural terms. But I want to say just two things. One is a lot of current theorizing entirely emphasizes classificatory processes, but classification of what? This goes back to our original discussion about materialism; I'm still anxious whether the constructionist position is just a new idealism in which we talk about how classificatory schemes relate to classificatory schemes. Whereas I think there are foundational issues, in terms of demography for example, that are not just a question of classification, it still seems to me that neither anthropology nor sociology is addressing the foundational issues of either ontology or materialism. This comes out fundamentally in recent discussions of the differences between genders, sex, sexuality and social roles.

Much recent writing has looked at ways in which sex differences are used as classificatory schemes for a range of social and cultural activities. The language of sexuality is used in terms of sacred and profane, strong and weak, hot and cold and so forth. Conventional distinctions between male and female have been made very problematic by both anthropology and sociology. But even if sexuality is produced by classificatory systems, it still seems to me that male and female bodies are organically, physiologically, biochemically different phenomena. I know there are problems in classifying biological sex differences. Biological difference is socially produced by the endless reproduction of human beings, but the classificatory systems can be seen as reflections upon differences in natural phenomena.

RF   Let me ask you about the politics of your position, because you have emphasized that your foundationalism partly springs from intellectual concerns but just as importantly reflects worries you have about the relevance of sociological or anthropological thought. You fear that if the disciplines entirely let go the foundationalism that supports a common human ontology this would prejudice your political concerns which, in a broad sense, are concerns about the types of intervention sociologists can make in life generally.

BT   I think there are two issues: one is the fundamental relationship between our embodiment and our identities, so that to change our embodiment is to change our identity. For example, mastectomy, major surgery and major interventions in the body actually change the nature of the personality and one's self-respect. The other is the idea of human frailty, that being embodied means we are frail in time and thus in relationship to our personalities. I want to retain the foundationalist view of that frailty of the human body in order to have a politics, on the one hand, of ecology and, on the other hand, of human rights. Human bodies need the institutional protection of citizenship and a political system that is sensitive to male/female differences, to questions about ageing and childhood and so on. There is a political logic behind retaining a foundationalist view in order to mount arguments about such human rights and to bring on to the agenda of sociology questions about emotion in relation to reason, and to make gender issues more central to sociological theory. The sociology of the body is part of an attack on a particular Cartesian, utilitarian paradigm of social relations, and an attempt to establish the basis of political action.

RF We could profitably explore the relation between political need and philosophical justification in relation to embodiment much further. Perhaps for now we could agree that a shared discourse of bodily ontology might at least be the legitimate objective – and probably a requirement – of a certain kind of political project. These issues bring us to the relationship of your work to feminist theory. To start with an obvious issue, how important to your work is the distinction you made between patrism and patriarchy in *The Body and Society*? That contrast does not appear prominently in your later work.

BT Whereas, in autobiographical terms, *Religion and Social Theory* responded to my religious background, *The Body and Society* was in part a consequence of my migration to Australia, where I became far more interested in and exposed to feminist social theory. I found the feminist critique of distinctions between convention/culture/nature useful in trying to think about the body. Feminism was clearly an important challenge, both politically and intellectually, to Marxism and sociology; it was important to engage with the issues raised by the feminist movement. One theme common to the three books is the examination of the various institutional orders (religion, law and medicine) which have controlled female reproduction, sexuality, child-care and family life. I have been sympathetic to writers like Talcott Parsons who, in the concept of the sick role, almost unwittingly developed a theory of social control.

However, I was less sympathetic to versions of radical feminism which argued that the position of women had not changed, and that the concept of patriarchy could be used to describe the social position of women from Constantine to late capitalism. In fact, this view of patriarchy came close to ascribing an essentialist character to sexual difference. Two things occurred to me as being important in trying to understand patriarchy. The first was that if you go back to the original use of the term by writers like Filmer then patriarchal relations are relations of hierarchical authority exercised over subordinate men and women. These relations were domestic but, since the royal household was public, state hierarchy had its origins in the King as father. You find similar ideas and practices in Islam. While Foucault does not, as far as I am aware, use the word patriarchy, his description of patriarchal relations between men in classical Greek homosexuality is brilliant. Secondly, I felt that the collapse of Christian authority in the West, legislation against sexual discrimination, programmes of positive discrimination and changes in the structure of the household did constitute 'social change'. I wanted to propose a new concept to describe these relations, namely a change from patriarchal to patristic authority. I was not suggesting naively that the position of women was in all respects necessarily better, but it did seem to me to be different.

RF Having made the critical point, did the usefulness of the distinction disappear?

BT Not at all. *Medical Power and Social Knowledge* was based on the argument that the patriarchal powers of the priest have been reassembled and secularized under the patristic authority of doctors. Hence, in the medical book I concentrated rather heavily on 'women's complaints'. Some of these ideas were first explored with Mike Hepworth in the research we did on confession (Hepworth and Turner 1982), where we looked for the origins of psychiatric practice in the traditional confessional. I have also done research in Australia, motivated by feminist critiques of medical professionalism, on how nurses complain about the gender division in medical practice (Turner 1986b). The problematic nature of anorexia nervosa as a 'condition' has always interested me as a research topic. I have taken a critical but sympathetic view of feminist research in this area of so-called eating disorders (Turner 1991a).

RF I detect, however, some important differences in your treatment of gender in these three books. Again, the emphasis upon a materialist perspective seems to diminish over time.

BT That is true. When I started this work, I wanted to show that changes in the economic organization of society, especially changes in the nature of property ownership, had major consequences for marriage practices and sexual mores. In fact, this argument first appeared in the discussion of medieval family and marriage customs in *The Dominant Ideology Thesis* (Abercrombie *et al.* 1980: 87–94). The basic idea of this study of ideology was that, as property has become in various ways 'de-personalized' with the decline of competitive capitalism, then the link between property and sexual regulation was broken. The argument, which has obviously been the target of much criticism, was that the economic structure of capitalism did not require a *specific* sexual ideology or a normative set of sexual practices. However, as my work developed, there was much less emphasis on this economic sociology, because Foucault's treatment of power/knowledge appeared far more interesting and relevant, especially to medical practice. In recent years, I have been increasingly influenced by writers like Emily Martin, Nancy Fraser, Donna Haraway, Elizabeth Wilson, Angela McRobbie, and others. The result is a shift away from my rather narrow preoccupations of the 1970s.

RF How do you think you will develop these themes?

BT While my early work was clearly influenced by feminism, a criticism made of it by my colleague Ken Plummer is that I have neglected gay literature and, in particular, I ignored the literature by writers like Jeff Weeks. I am now very conscious of the substantial and growing gay literature on the body and sexuality. However, one can think of other lacunae I should examine: comparative anthropology, art history, the history of science, and the sociology of AIDS. One would also imagine that there will be an explosion of studies broadly around the postmodern body. I would predict a fruitful theoretical inter-action between feminism, gay studies, the sociology of the body and ecologism. The postmodernist critique of instrumental rationality provides an obvious linkage between green politics and the sociology of gender.

RF Why have you previously neglected this material?

BT I think there are two reasons. The first is that I have been involved in classical sociology and in fundamental questions about the theory of social action in Marx, Weber, Parsons and Simmel. My assumption was that if one could resolve basic issues in the nature of social action and social structure, then substantive issues in the sub-fields of sociology would fall into place. Secondly, I have become specifically concerned with teaching and research in medical sociology, where the issue of the body appears to have a 'natural' place.

However, my fundamental concern is with a subject which we touched on earlier, namely the relationship between politics and the body. If I can retain a foundationalist epistemology of the body along the lines I have already indicated, I believe it is possible to move philosophically from a view of human frailty as the basis for a global *conditio humana* (perhaps a secular parallel to Dietrich Bonhoeffer's *sanctorum communio*) to a theory of human rights as an extension of modern citizenship. Somewhat against the trend of contemporary theory, I want an overall theoretical and practical coherence linking my work in the sociology of religion, medical sociology and political sociology. This might be a good place to end. At the beginning of our discussion you asked me whether my work had an overall goal. Perhaps not, but I am aware of a politico-moral objective, which is bound up with my view that any sociology worth doing engages us in moral and political activity.

REFERENCES

Abercrombie, N., S. Hill and B.S. Turner (1980) *The Dominant Ideology Thesis*, London: Allen & Unwin.

Alexander, J. (1987) *Twenty Lectures, Sociological Theory since World War II*, New York: Columbia University Press.

Althusser, L. (1969) *For Marx*, London: Allen Lane.

Dreyfus, H.L. and P. Rabinow (1986) *Michel Foucault, Beyond Structuralism and Hermeneutics*, Brighton: Harvester Press.

Durkheim, E. (1960) *The Division of Labour in Society*, Glencoe, Ill: Free Press.

Durkheim, E. and M. Mauss (1963) *Primitive Classification*, Chicago: Chicago University Press.

Featherstone M. and M. Hepworth (1991) 'The mask of ageing and the postmodern life course', pp. 371–89 in M. Featherstone, M. Hepworth and B.S. Turner (eds) *The Body, Social Process and Cultural Theory*, London: Sage.

Featherstone, M., M. Hepworth and B.S. Turner (eds) (1991) *The Body, Social Process and Cultural Theory*, London: Sage.

Foucault, M. (1974a) *The Order of Things, an Archaeology of the Human Sciences*, London: Tavistock.

Foucault, M. (1974b) *The Archaeology of Knowledge*, London: Tavistock.

Foucault, M. (1977) *Discipline and Punish, the Birth of the Prison*, London: Tavistock.

Foucault, M. (1980) (C. Gordon, ed.) *Michel Foucault, Power/Knowledge*, Brighton: Harvester Press.

Frank, A.W. (1991) 'From sick role to health role: deconstructing Parsons', pp. 205–16 in R. Roberston and B.S. Turner (eds) *Talcott Parsons, Theorist of Modernity*, London: Sage.

Giddens, A. (1984) *The Constitution of Society. Outline of the Theory of Structuration*, Cambridge: Polity Press.

Heald, S. (1989) *Controlling Anger, the Sociology of Gisu Violence*, Manchester: Manchester University Press.

Hennis, W. (1988) *Max Weber, Essays in Reconstruction*, London: Allen & Unwin.

Hepworth, M. and B.S. Turner (1982) *Confession, Studies in Deviance and Religion*, London: Routledge & Kegan Paul.

Hertz, R. (1960) *Death and the Right Hand*, London: Cohen & West.

Holton, R.J. and B.S. Turner (1989) *Max Weber on Economy and Society*, London: Routledge.

Jameson, F. (1973) 'The vanishing mediator: narrative structure in Max Weber', *New German Critique* vol. 1:52–89.

Liebersohn, H. (1988) *Fate and Utopia in German Sociology 1870–1923*, Cambridge, Mass: MIT Press.

Martin, L.H., H. Gutman and P.H. Hutton (eds) (1988) *Technologies of the Self, a Seminar with Michel Foucault*, London: Tavistock.

Mauss, M. (1979) *Sociology and Psychology*, London: Routledge & Kegan Paul.

Needham, R. (1981) *Circumstantial Deliveries*, Berkeley: University of California Press.

Parsons, T. (1951) *The Social System*, London: Routledge & Kegan Paul.

Robertson, R. (1974) 'Towards the identification of the major axes of sociological analysis', pp. 107–24 in J. Rex (ed.) *Approaches to Sociology, an Introduction to Major Trends in British Sociology*, London: Routledge & Kegan Paul.

Robertson, R. and B.S. Turner (eds) (1991) *Talcott Parsons, Theorist of Modernity*, London: Sage.

Rosaldo, M.Z. (1980) *Knowledge and Passion, Ilongot Notions of Self and Society*, Cambridge: Cambridge University Press.

Stauth, G. and B.S. Turner (1988) *Nietzsche's Dance, Resentment, Reciprocity and Resistance in Social Life*, Oxford: Basil Blackwell.

Turner, B.S. (1974) *Weber and Islam, a Critical Study*, London: Routledge & Kegan Paul.

Turner, B.S. (1981) *For Weber, Essays on the Sociology of Fate*, London: Routledge & Kegan Paul.

Turner, B.S. (1983) *Religion and Social Theory*, London: Heinemann Educational Books.

Turner, B.S. (1984) *The Body and Society, Explorations in Social Theory*, Oxford: Basil Blackwell.

Turner, B.S. (1985) 'The practices of rationality, Michel Foucault, medical history and sociological theory', pp. 193–213 in R. Fardon (ed.) *Power and Knowledge, Anthropological and Sociological Approaches*, Edinburgh: Scottish Academic Press.

Turner, B.S. (1986a) *Equality*, London: Tavistock.

Turner, B.S. (1986b) *Citizenship and Capitalism, the Debate over Reformism*, London: Allen & Unwin.

Turner, B.S. (1987) *Medical Power and Social Knowledge*, London: Sage.

Turner, B.S. (1988) 'Some reflections on cumulative theorizing in sociology', pp. 131–47 in J.H. Turner (ed.) *Theory Building in Sociology*, Newbury Park: Sage.

Turner, B.S. (1991a) *Religion and Social Theory*, London: Sage (second edition, with a new introduction).

Turner, B.S. (1991b) 'Recent developments in the theory of the body', pp. 1–35 in M. Featherstone, M. Hepworth and B.S. Turner (eds) *The Body, Social Process and Cultural Theory*, London: Sage.

Turner, B.S. (1992) *Max Weber, from History to Modernity*, London: Routledge.

Appendix
Bryan S. Turner's publications on the sociology of the body and medical sociology

1 'The body and religion, towards an alliance of medical sociology and the sociology of religion', *Annual Review of the Social Sciences of Religion* 1980, vol. 4:247–86.

2 'Weber on medicine and religion', pp. 177–99 in Bryan S. Turner *For Weber, Essays on the Sociology of Fate*, London: Routledge & Kegan Paul, 1981.

3 'The discourse of diet', *Theory, Culture & Society* 1982a, vol. 1(1):23–32.

4 'The government of the body, medical regimens and the rationalisation of diet', *The British Journal of Sociology* 1982b, vol. 33:254–69.

5 'Secular bodies and the dance of death', pp. 227–41 in Bryan S. Turner *Religion and Social Theory, a Materialist Perspective*, London: Heinemann, 1983.

6 *The Body and Society, Explorations in Social Theory*, Oxford: Basil Blackwell, 1984.

7 'More on "The government of the body": a reply to Naomi Aronson', *The British Journal of Sociology* 1985a, vol. 36: 151–4.

8 'The practices of rationality, Michel Foucault, medical history and sociological theory', pp. 193–213 in Richard Fardon (ed.) *Power and Knowledge, Anthropological and Sociological Approaches*, Edinburgh: Scottish Academic Press, 1985b.

9 'Knowledge, skill and occupational strategies, the professionalisation of paramedical groups', *Community Health Studies* 1985c, vol. 9:38–47.

10 'Sociology of the body', pp. 77–8 in A. Kuper and J. Kuper (eds) *The Social Science Encyclopedia*, London: Routledge & Kegan Paul, 1985d.

11 'Sickness and social structure: Parsons's contribution to medical sociology', pp. 107–42 in R.J. Holton and B.S. Turner *Talcott Parsons on Economy and Society*, London: Routledge & Kegan Paul, 1986a.

12 'The vocabulary of complaints, nursing, professionalism and job context', *The Australian and New Zealand Journal of Sociology* 1986b, vol. 22(3):368–86.

13 'The rationalization of the body: reflections on modernity and discipline', pp. 222–41 in S. Whimster and S. Lash (eds) *Max Weber, Rationality and Modernity*, London: Allen & Unwin, 1987a.

14 *Medical Power and Social Knowledge*, London: Sage, 1987b.

15 'Agency and structure in the sociology of sickness', in I. Pilowsky (ed.) *Psychiatric Medicine* 1987c, vol. 5(1):29–37.

16 'A note on nostalgia', *Theory, Culture & Society* 1987d, vol. 4(1):147–56.

17 'Of the despisers of the body – Nietzsche and French social theory', pp. 180–208 in G. Stauth and B.S. Turner *Nietzsche's Dance, Resentment, Reciprocity and Resistance in Social Life*, Oxford: Basil Blackwell, 1988.

18 'Ageing, status politics and sociological theory', *The British Journal of Sociology* 1989a, vol. 40(2):588–606.

19 *El Cuerpo y la Sociedad, exploraciones en teoria*, Mexico: Fondo de Cultura Economica (Spanish translation of No. 6).

20 'The talking disease: Hilda Bruch and anorexia nervosa', *The Australian and New Zealand Journal of Sociology* 1990a, vol. 26: 157–69.

21 'The anatomy lesson: a note on the Merton thesis', *The Sociological Review* 1990b, vol. 38(1):1–18.

22 'The interdisciplinary curriculum, from social medicine to postmodernism', *Sociology of Health and Illness* 1990c, vol. 12(1):1–23.

23 'Recent developments in the theory of the body', pp. 1–35 in M. Featherstone, M. Hepworth and B.S. Turner (eds) *The Body, Social Process and Cultural Theory*, London: Sage, 1991a.

24 'Missing bodies, towards a sociology of embodiment', *Sociology of Health and Illness* 1991b, vol. 13:265–72.

25 'Patriarchy and anorexia nervosa: a reply to Jan Horsfall', *The Australian and New Zealand Journal of Sociology* 1991, vol. 27(2):235–8.

26 *Kroppen i Samfundet, Teorier om krop og kultur*, Copenhagen: Hans Reitzels Forlag, 1992 (Danish translation of No. 6).

Name index

Subject index